Food Innovation as a Means of Developing Healthier and More Sustainable Foods

Food Innovation as a Means of Developing Healthier and More Sustainable Foods

Editors

Adrián Rabadán
Rodolfo Bernabéu

MDPI • Basel • Beijing • Wuhan • Barcelona • Belgrade • Manchester • Tokyo • Cluj • Tianjin

Editors

Adrián Rabadán
Higher Technical School of Agricultural and Forestry Engineering, University of Castilla-La Mancha
Spain

Rodolfo Bernabéu
Higher Technical School of Agricultural and Forestry Engineering, University of Castilla-La Mancha
Spain

Editorial Office
MDPI
St. Alban-Anlage 66
4052 Basel, Switzerland

This is a reprint of articles from the Special Issue published online in the open access journal *Foods* (ISSN 2304-8158) (available at: https://www.mdpi.com/journal/foods/special_issues/innovation_food).

For citation purposes, cite each article independently as indicated on the article page online and as indicated below:

LastName, A.A.; LastName, B.B.; LastName, C.C. Article Title. *Journal Name* **Year**, *Volume Number*, Page Range.

ISBN 978-3-0365-2089-6 (Hbk)
ISBN 978-3-0365-2090-2 (PDF)

Contents

foods

MDPI

Editorial

Food Innovation as a Means of Developing Healthier and More Sustainable Foods

Adrián Rabadán *, Roberto Nieto and Rodolfo Bernabéu

Higher Technical School of Agriculture and Forestry Engineering, University of Castilla-La Mancha, 02071 Albacete, Spain; Roberto.nieto@alu.uclm.es (R.N.); Rodolfo.Bernabeu@uclm.es (R.B.)
* Correspondence: adrian.rabadan@uclm.es

Abstract: The current demand for healthy and sustainable foods has encouraged the development of new alternatives even in traditional products. Improved foods may be produced by reducing the amount of some ingredients, adding new ones, or replacing traditionally used ingredients for others. By reformulating their products, manufacturers can offer healthier choices for an ever-growing number of consumers interested in maintaining a balanced diet. In addition, the market demand for more sustainable foods contributes to a lower environmental impact in their production. In this regard, current areas of interest include the production of foods using a lower number of inputs, as well as the utilization of food by-products, to improve the amount and quality of available foods. Another aspect to be considered is that not all consumers are willing to eat foods produced with new ingredients or novel technologies. Hence, the development of innovations in food products should take into account the influence of so-called "consumer food neophobia".

Keywords: novel foods; functional food; food by-product; sustainability; food neophobia

Citation: Rabadán, A.; Nieto, R.; Bernabéu, R. Food Innovation as a Means of Developing Healthier and More Sustainable Foods. *Foods* **2021**, *10*, 2069. https://doi.org/10.3390/foods10092069

Received: 19 August 2021
Accepted: 24 August 2021
Published: 1 September 2021

Consumers are increasingly aware of the composition of their food and the way in which it is produced. These new demands have boosted food innovation. Within food innovation studies, two main trends have dominated the studies developed in Western countries in recent years [1]. The first is the connection between food and health, which has encouraged the production of health-promoting foods. The second is the study of the impact of food production on the environment and the identification of new strategies (e.g., meat replacement) in order to reduce this impact.

Although an increasing number of consumers are open to trying novel foods [2], food innovation may face rejection from consumers that demand more natural food or foods produced using traditional production schemes [3]. This is referred to as consumer neophobia, which has been defined as the reluctance of consumers to try new or unfamiliar foods. This expected rejection of novel foods is one of the main issues that the implementation of food innovation strategies will have to face in the near future.

The current trend of innovations designed to develop more healthy and sustainable foods includes a wide range of very different approaches. Two approaches that have received major attention are the substitution of food components and the development of new production technologies. In this regard, new starches and other hydrocolloids have expanded the use and substitution of many ingredients [4], and traditional cereal flours have been replaced by other flours such as pulse flours [5] or nut flours [6,7]. New production methods include, for example, the use of critical CO_2 to conduct the extraction of different components [8] and the use of ultrasound assisted extraction or microwave assisted extraction [9,10]. From the point of view of consumers, studies have focused on consumers' attitudes towards organic products [11], gluten-free products [6], and the various kinds of innovations that result in the production of more natural or more sustainable foods [3,12].

This Special Issue provides a multidisciplinary view of food innovation including studies focused on the production of healthier and more sustainable foods by using novel ingredients, food by-products, or new food production processes. Additionally, studies about consumers' perceptions of food innovation or innovative foods have also been considered.

Rábago-Panduro et al. [13] investigated the effect of pulsed electric fields on the extraction yield and stability of oils obtained from pecan nuts (*Carya illinoinensis* (Wangenh. K. Koch)). In their study, the authors evaluated how three specific energy inputs affected oil extraction yield, oil stability, the microstructure of kernels and their by-products' phenolic compounds and antioxidant capacity. The results show that the use of pulsed electric fields does not improve oil extraction yield, showing that factors such as the moisture or microstructure of kernels play a key role in their effectiveness.

Gómez-Saéz et al. [14] evaluated the effect of three concentrations of saffron on vacuum-packed dry-cured ham. In this study, the authors evaluated several factors, including pH, color, sensorial quality and the safranal content of ham, at 0, 7, 14, 28, and 60 days of storage. Their results show that the addition of saffron did not affect the pH or the color of ham stored for 28 days. However, the storage period affected pH values with a decline observed from day 28. Regarding sensory analysis, significant differences were found in visual appearance, flavor, and odor at different days of storage. The safranal content also showed variations with the time at days 14 and 60.

Rabadán [15] studied consumers' attitudes towards wine innovation using a sample of 400 Spanish wine consumers using several scales, including the "Wine Neophobia Scale". His results show that four different segments of consumers exist, showing different attitudes towards technological innovation in the wine sector. The group that shows more positive attitudes towards technological innovation (product and process innovation) is formed by consumers with the highest incomes and levels of education. With that in mind, he concludes that innovations in the wine sector should be directed to consumers within that segment.

In order to improve the nutritional and sensory characteristics of cookies, Martínez et al. [6] studied the effect of the substitution of wheat flour by seed defatted flours in cookies' elaboration. Four different defatted flours were used: flax, sesame, chia, and poppy. The results show that the use of defatted seed flours resulted in cookies with a higher protein content and a lower content of carbohydrates. Regarding sensory analysis, sesame and flax cookies obtained similar values than traditional cookies, while cookies elaborated with chia and poppy flours received the least positive evaluations.

Hinestroza-Córdoba et al. [16] evaluated the use of vacuum impregnation and high-pressure homogenization in order to obtain functional ingredients from lulo fruit (*Solanum quitoense* Lam.). Specifically, they studied the physicochemical and antioxidant properties of the lulo fruit and its juice. In their results, they found differences between the fruit and the juice and concluded that high-pressure homogenization increases the antiradical capacity of the juice and the diversity of polyphenols.

Zhang et al. [17] studied the willingness of consumers to pay for enhanced mandatory labeling of genetically modified soybean oil using a large sample of Chinese consumers. Their results show that Chinese consumers show positive attitudes towards traceability codes, reporting a willingness to pay of RMB 8.92 and for the labelling of allergen presences (RMB 6.57). In their conclusions, they state that policy strategies for enhanced mandatory labelling in genetically modified soybean oil will benefit consumers.

Brugarolas et al. [18] analyzed the behavior of consumers during the COVID-19 pandemic in order to create innovative strategies for the agri-food sector to cope with this new scenario. They found that 61% of consumers modified their buying behavior during lockdown, with increasing food stockpiling being the most common observed change. Regarding the degree of change in buying behavior, four different consumer segments were identified. The authors propose new strategies that companies could implement in order to deal with this new scenario, including the creation of a larger stock of non-perishable foods, increasing production capabilities, or boosting online sales.

Jiménez-Ortega et al. [19] evaluated the health-promoting properties of crocetin in order to promote its consideration as a healthy natural colorant. Specifically, they evaluated the ability of crocetin to reduce lipid accumulation during the differentiation of 3T3-L1 preadipocytes. In their results they state that 5 μM of crocetin decreased intracellular fat by 22.6% without affecting viability or lipid droplet generation. This result encourages the use of crocetin in dietary therapies intended to reverse adipose tissue accumulation in obesity cases.

Rabadán et al. [20] studied the effect of a melon cultivar (*Cucumis melo* L.) and an oil extraction method on the nutritional quality of melon seed oil. Nine different melon cultivars and two oil extraction methods (hydraulic and screw press) were evaluated. The results show that higher oil yields were obtained using the screw press. However, oils obtained with the hydraulic press resulted in higher quality oils. The obtained oils showed significant differences in their linoleic (51–69%) and oleic (15–34%) acids content. Vitamin E content also showed significant differences among cultivars. γ-tocopherol was the main isoform found in oils (range 99.81–456.73 mg/kg), followed by α- and δ-tocopherols. The principal-component analysis concluded that cultivars *Honew Dew* and *Blanco de Ribatejo* show the most promising characteristics to produce high-quality melon seed oils.

Roncero et al. [21] presented a review about almond kernel composition, making specific reference to the non-lipid components that traditionally have received less attention in the literature. They conclude that almonds are rich in proteins (8–35%), including important concentrations of the globulin-albumin fraction. The carbohydrate fraction (14–28%) is mainly compound by soluble sugars (such as sucrose) and starch. Regarding mineral elements, relevant concentrations of potassium and phosphorus were reported.

All of the above studies result in a wide multidisciplinary approach about the current state of the art in food innovation. The development of novel ingredients, foods, and innovative food production technologies, while simultaneously considering consumers' perception of food innovation, will be crucial in order to produce healthier, safer, and more sustainable foods in the future.

Author Contributions: Conceptualization, A.R.; writing—original draft preparation, A.R., R.N., R.B.; writing—review and editing, A.R., R.B.; supervision, A.R., R.B.; project administration, A.R. All authors have read and agreed to the published version of the manuscript.

Funding: This research received no external funding.

Acknowledgments: The Guest Editors thank all the colleagues for their valuable contributions in this Special Issue. The Guest Editors would also like to thank the editorial staff of MDPI for their support.

Conflicts of Interest: The authors declare no conflict of interest.

References

1. Tuorila, H.; Hartmann, C. Consumer responses to novel and unfamiliar foods. *Curr. Opin. Food Sci.* **2020**, *33*, 1–8. [CrossRef]
2. Rabadán, A.; Bernabéu, R. A systematic review of studies using the Food Neophobia Scale: Conclusions from thirty years of studies. *Food Qual. Prefer.* **2021**, *93*, 104241. [CrossRef]
3. Rabadán, A.; Bernabéu, R. An approach to eco-innovation in wine production from a consumer's perspective. *J. Clean. Prod.* **2021**, *310*, 127479. [CrossRef]
4. Porwal, V.B.; Bharath Kumar, S.; Madhumathi, R.; Prabhasankar, P. Influence of health based ingredient and its hydrocolloid blends on noodle processing. *J. Food Meas. Charact.* **2014**, *8*, 283–295. [CrossRef]
5. Lazo-Vélez, M.A.; Garzon, R.; Guardado-Félix, D.; Serna-Saldivar, S.O.; Rosell, C.M. Selenized chickpea sourdoughs for the enrichment of breads. *LWT* **2021**, *150*, 112082. [CrossRef]
6. Martínez, E.; García-Martínez, R.; Álvarez-Ortí, M.; Rabadán, A.; Pardo-Giménez, A.; Pardo, J.E. Elaboration of Gluten-Free Cookies with Defatted Seed Flours: Effects on Technological, Nutritional, and Consumer Aspects. *Foods* **2021**, *10*, 1213. [CrossRef] [PubMed]

7. Rabadán, A.; Álvarez-Ortí, M.; Martínez, E.; Pardo-Giménez, A.; Zied, D.C.; Pardo, J.E. Effect of replacing traditional ingredients for oils and flours from nuts and seeds on the characteristics and consumer preferences of lamb meat burgers. *LWT* **2021**, *136*, 110307. [CrossRef]
8. Vidović, S.; Tomšik, A.; Vladić, J.; Jokić, S.; Aladić, K.; Pastor, K.; Jerković, I. Supercritical Carbon Dioxide Extraction of Allium ursinum: Impact of Temperature and Pressure on the Extracts Chemical Profile. *Chem. Biodivers.* **2021**, *18*, e2100058. [CrossRef]
9. Han, L.; Cao, S.; Yu, Y.; Xu, X.; Cao, X.; Chen, W. Modification in physicochemical, structural and digestive properties of pea starch during heat-moisture process assisted by pre- and post-treatment of ultrasound. *Food Chem.* **2021**, *360*, 129929. [CrossRef]
10. Boggia, R.; Turrini, F.; Villa, C.; Lacapra, C.; Zunin, P.; Parodi, B. Green extraction from pomegranate marcs for the production of functional foods and cosmetics. *Pharmaceuticals* **2016**, *9*, 63. [CrossRef] [PubMed]
11. Yormirzoev, M.; Li, T.; Teuber, R. Consumers' willingness to pay for organic versus all-natural milk—Does certification make a difference? *Int. J. Consum. Stud.* **2021**, *45*, 1020–1029. [CrossRef]
12. Lockie, S.; Lyons, K.; Lawrence, G.; Mummery, K. Eating 'Green': Motivations behind organic food consumption in Australia. *Sociol. Rural.* **2002**, *42*, 23–40. [CrossRef]
13. Rábago-Panduro, L.M.; Morales-de la Peña, M.; Romero-Fabregat, M.P.; Martín-Belloso, O.; Welti-Chanes, J. Effect of Pulsed Electric Fields (PEF) on Extraction Yield and Stability of Oil Obtained from Dry Pecan Nuts (*Carya illinoinensis* (Wangenh. K. Koch)). *Foods* **2021**, *10*, 1541. [CrossRef]
14. Gómez-Sáez, E.M.; Moratalla-López, N.; Alonso, G.L.; Vergara, H. Partial Characterization of the Impact of Saffron on the Sensory and Physicochemical Quality Traits of Dry-Cured Ham. *Foods* **2021**, *10*, 1506. [CrossRef]
15. Rabadán, A. Consumer Attitudes towards Technological Innovation in a Traditional Food Product: The Case of Wine. *Foods* **2021**, *10*, 1363. [CrossRef]
16. Hinestroza-Córdoba, L.I.; Barrera, C.; Seguí, L.; Betoret, N. Potential Use of Vacuum Impregnation and High-Pressure Homogenization to Obtain Functional Products from Lulo Fruit (*Solanum quitoense Lam.*). *Foods* **2021**, *10*, 817. [CrossRef]
17. Zhang, M.; Fan, Y.; Cao, J.; Chen, L.; Chen, C. Willingness to Pay for Enhanced Mandatory Labelling of Genetically Modified Soybean Oil: Evidence from a Choice Experiment in China. *Foods* **2021**, *10*, 736. [CrossRef] [PubMed]
18. Brugarolas, M.; Martínez-Carrasco, L.; Rabadán, A.; Bernabéu, R. Innovation Strategies of the Spanish Agri-Food Sector in Response to the Black Swan COVID-19 Pandemic. *Foods* **2020**, *9*, 1821. [CrossRef]
19. Jiménez-Ortega, E.; Braza-Boïls, A.; Burgos, M.; Moratalla-López, N.; Vicente, M.; Alonso, G.L.; Nava, E.; Llorens, S. Crocetin Isolated from the Natural Food Colorant Saffron Reduces Intracellular Fat in 3T3-L1 Adipocytes. *Foods* **2020**, *9*, 1648. [CrossRef] [PubMed]
20. Rabadán, A.; Nunes, M.A.; Bessada, S.M.F.; Pardo, J.E.; Oliveira, M.B.P.P.; Álvarez-Ortí, M. From By-Product to the Food Chain: Melon (*Cucumis melo* L.) Seeds as Potential Source for Oils. *Foods* **2020**, *9*, 1341. [CrossRef] [PubMed]
21. Roncero, J.M.; Álvarez-Ortí, M.; Pardo-Giménez, A.; Rabadán, A.; Pardo, J.E. Review about Non-Lipid Components and Minor Fat-Soluble Bioactive Compounds of Almond Kernel. *Foods* **2020**, *9*, 1646. [CrossRef] [PubMed]

foods

MDPI

Article

Effect of Pulsed Electric Fields (PEF) on Extraction Yield and Stability of Oil Obtained from Dry Pecan Nuts (*Carya illinoinensis* (Wangenh. K. Koch))

Lourdes Melisa Rábago-Panduro [1,2], Mariana Morales-de la Peña [3], María Paz Romero-Fabregat [2], Olga Martín-Belloso [1,2] and Jorge Welti-Chanes [1,*]

1 FEMSA Biotechnology Center, Tecnológico de Monterrey, School of Engineering and Sciences, Eugenio Garza Sada Ave. 2501, 64849 Monterrey, Mexico; a00821713@itesm.mx (L.M.R.-P.); olga.martin@udl.cat (O.M.-B.)
2 AGROTECNIO CERCA Center, Department of Food Technology, University of Lleida, Rovira Roure Ave. 191, 25198 Lleida, Spain; mariapaz.romero@udl.cat
3 Bioengineering Center, Tecnológico de Monterrey, School of Engineering and Sciences, Epigmenio González Ave. 500, 76130 Queretaro, Mexico; mariana.moralesdlp@tec.mx
* Correspondence: jwelti@tec.mx; Tel.: +52-81-83284262

Citation: Rábago-Panduro, L.M.; Morales-de la Peña, M.; Romero-Fabregat, M.P.; Martín-Belloso, O.; Welti-Chanes, J. Effect of Pulsed Electric Fields (PEF) on Extraction Yield and Stability of Oil Obtained from Dry Pecan Nuts (*Carya illinoinensis* (Wangenh. K. Koch)). *Foods* **2021**, *10*, 1541. https://doi.org/10.3390/foods10071541

Academic Editors: Adrián Rabadán and Rodolfo Bernabéu

Received: 30 May 2021
Accepted: 30 June 2021
Published: 3 July 2021

Abstract: Pulsed electric fields (PEF) have been reported to increase the total oil extraction yield (OEY_{TOTAL}) of fresh pecan nuts maintaining oil characteristics and increasing phenolic compounds in the remaining by-product. However, there is no information regarding the PEF effect on dry pecan nuts. Dry kernels were pretreated at three specific energy inputs (0.8, 7.8 and 15.0 kJ/kg) and compared against untreated kernels and kernels soaked at 3, 20 and 35 min. OEY_{TOTAL}, kernels microstructure, oil stability (acidity, antioxidant capacity (AC), oil stability index, phytosterols and lipoxygenase activity), along with by-products phenolic compounds (total phenolics (TP), condensed tannins (CT)) and AC were evaluated. Untreated kernels yielded 88.7 ± 3.0%, whereas OEY_{TOTAL} of soaked and PEF-treated kernels were 76.5–83.0 and 79.8–85.0%, respectively. Kernels microstructural analysis evidenced that the 0.8 kJ/kg pretreatment induced oleosomes fusion, while no differences were observed in the stability of extracted oils. PEF applied at 0.8 kJ/kg also increased by-products CT by 27.0–43.5% and AC by 21.8–24.3% compared to soaked and untreated kernels. These results showed that PEF does not improve OEY_{TOTAL} when it is applied to dry pecan nuts, demonstrating that kernels' moisture, oil content and microstructure play an important role in the effectiveness of PEF.

Keywords: pulsed electric fields; pecan nut oil; oil extraction yield; microstructural analysis; oil stability; enzyme activity

1. Introduction

Pulsed electric fields (PEF) involved the application of intermittent electric fields of varying intensity (0.1–50 kV/cm) and short duration (μs–ms) [1]. Currently, this technology has been used as a pretreatment to oil extraction from rapeseeds, sunflower seeds, sesame seeds and olives enhancing oil extraction yield (OEY) and increasing oil acidity [2–5]. The improvement of OEY has been attributed to the electroporation phenomenon, while the increase of oil acidity was associated with triacylglycerols hydrolysis by lipase activity. In a previous study, PEF were applied to fresh pecan nut kernels as a pretreatment employing different specific energy inputs [5]. Results indicated that PEF-treated kernels yielded 21.4% more oil than untreated kernels. Nonetheless, pecan nuts are usually submitted to a drying process for commercialization and the effect of this technology on the OEY of dry kernels has not been reported yet. Furthermore, studies concerning the PEF effect on seeds microstructure and oil acidity are scarce.

Pecan nut kernels (*Carya illinoinensis* (Wangenh. K. Koch)) are well-known for their high concentration of phenolic compounds, mono- and polyunsaturated fatty acids, phytosterols, and tocopherols [6,7]. Besides, pecan nut oil differentiates from other tree nut oils due to its content of polyunsaturated fatty acids (PUFA), β-sitosterol and γ-tocopherol [8,9]. Thus, to extend kernels′ shelf-life and avoid oil oxidation, a drying process is carried out reducing the moisture content (≤5.5%) and modifying kernels microstructure [10,11]. In pecan nuts, fatty acids are stored as triacylglycerols within oleosomes. These organelles are formed by a monolayer of phospholipids, proteins, and enzymes like lipoxygenase that is the key enzyme in lipids oxidation [12–15].

During oxidative deterioration, lipids are oxidized to hydroperoxides and then to secondary oxidation products that increment free radicals′ concentration. Hence, these compounds are considered important biomarkers to measure oil oxidative deterioration [16]. Acidity is the measurement of free fatty acids concentration produced by ester bonds hydrolysis of lipids by either enzyme action, heat or moisture [17]. Thus, its increment is related with a loss of oil stability. Another method to evaluate oil stability is the oil stability index (OSI). In the OSI method, oil is oxidized by air and high temperatures, producing volatile acids that are dissolved in water to follow the change in electrical conductivity [15]. The OSI value expresses the time necessary to complete oil oxidation at given experimental conditions being indicated by a drop of conductivity. Hence, a high OSI value implies a high resistance to oxidation of the analyzed oil [17].

This study was carried out to investigate the effect of PEF on OEY, OEY_{TOTAL} and microstructure of dry pecan nuts. In addition, the stability of the extracted oil was evaluated by determination of acidity, antioxidant capacity, OSI, phytosterols concentration and LOX activity as well as phenolic compounds and antioxidant capacity of the generated by-products hereinafter referred to as cakes.

2. Materials and Methods

2.1. Chemicals

Acetonitrile, ethyl acetate, methanol (MeOH), hexane, water HPLC grade (H_2O), β-mercaptoethanol (β-ME), 2,2-diphenyl-1-picrylhydrazyl (DPPH), boric acid (H_3BO_3), hydrochloric acid (HCl), glutaraldehyde, osmium tetroxide, polyvinylpolypyrrolidone (PVPP), potassium hydroxide solution (KOH 0.1 M), sodium acetate (CH_3COONa), sodium phosphate monobasic (NaH_2PO_4), sodium phosphate dibasic (Na_2HPO_4), Triton X-114 and uranyl acetate were acquired from Sigma-Aldrich (St. Louis, MO, USA) along with 6-hydroxy-2,5,7,8-tetramethylchromane-2-carboxylic acid (Trolox), linoleic acid, β-sitosterol, stigmasterol, campesterol, catechin, gallic acid and vanillin. Epoxy EMbed 812 resin was purchased from Electron Microscopy Sciences (Hatfield, PA, USA). Acetic acid (CH_3COOH), ethanol (EtOH), sodium carbonate (Na_2CO_3) and KOH were purchased from DEQ (Monterrey, Nuevo León, Mexico).

2.2. Pecan Nuts

Kernel halves of dry pecan nuts (*Carya illinoinensis*) from Alesto brand were purchased in a local market in Lleida (Spain). In this study, dry kernels were separated in three experimental groups comparing PEF-treated kernels against reference and control kernels. No soaking nor PEF processing was performed on reference kernels, while control kernels were placed in tap water (1:3 w/w) for 3, 20 and 35 min to assess the effect of soaking at each PEF pretreatment. Kernels pretreated by PEF were immersed in tap water and processed. After a 10 min draining, control and PEF-treated kernels (1.0 g) were separated for moisture and microstructural analysis. Before freeze-drying, kernel samples (2.0 g) were taken for determination of lipoxygenase (LOX) activity. The remaining kernels were frozen (−16 °C, 24 h), freeze-dried (−50 °C, 1 mbar, 72 h), and stored at −40 °C until oil extraction.

2.3. Pulsed Electric Fields Application

The application of PEF was performed in a batch-system using a 20 × 8 cm methacrylate container with stainless-steel parallel electrodes as the treatment chamber. The batch-system was equipped with a 0.1 μF capacitator (Physics International, USA), a pulse generator (PT-55, Pacific Atlantic Electronics Inc., El Cerrito, CA, USA) and a TG-70 gas control unit. Based on a previous study [18], dry kernels were immersed in tap water (1:3 *w*/*w*) to apply 10, 99 and 192 monopolar exponential-wave pulses at an electric field strength of 5.0 kV/cm. These electrical conditions equaled to specific energy inputs of 0.8, 7.8 and 15.0 kJ per kg of pecan nut kernels in wet basis (*W*, kJ/kg wb), respectively.

2.4. Oil Mechanical Extraction

Oil extraction was performed on freeze-dried kernels (85.0 g) employing a screw-type press (YD-ZY-02A, Yoda Europe, Hangzhou, Zhejiang, China). Pecan nut oil was stored in 50 mL centrifuge tubes flushing N_2 into the headspace and the generated cakes were placed in polyethylene bags and vacuum sealed. Oil and cake samples were stored at −40 °C until analyses.

Oil Extraction Yield

Reference, control and PEF-treated kernels oil extraction yields (OEY, %) were calculated as follows:

$$\mathrm{OEY} = \frac{(\mathrm{m_K} \times \mathrm{L_K}) - (\mathrm{m_C} \times \mathrm{L_C})}{(\mathrm{m_K} \times \mathrm{L_K})} \times 100, \quad (1)$$

where $\mathrm{m_K}$ and $\mathrm{L_K}$ are kernels mass (g) and oil content (g/100 g), and $\mathrm{m_C}$ and $\mathrm{L_C}$ are cakes mass and oil content, all in dry basis (db). Next, the oil extracted into the soaking water ($\mathrm{o_{SW}}$) of control and PEF-treated kernels was determined using the oil content of reference kernels ($\mathrm{L_{Reference}}$):

$$\mathrm{o_{SW}} = (\mathrm{m_K} \times \mathrm{L_{Reference}}) - (\mathrm{m_K} \times \mathrm{L_K}), \quad (2)$$

Total oil extraction yield ($\mathrm{OEY_{TOTAL}}$, %) of control and PEF-treated kernels was estimated using the $\mathrm{o_{SW}}$:

$$\mathrm{OEY_{TOTAL}} = \frac{[(\mathrm{m_K} \times \mathrm{L_K}) - (\mathrm{m_C} \times \mathrm{L_C})] + \mathrm{o_{SW}}}{(\mathrm{m_K} \times \mathrm{L_K})} \times 100, \quad (3)$$

2.5. Kernels Analysis

2.5.1. Moisture

The AOAC Official Method 920.151 [19] was used to analyze kernels moisture. Results were expressed as g per 100 g of kernels db (g/100 g db).

2.5.2. Oil Content

Oil content was analyzed as reported by Villarreal-Lozoya et al. [20]. Samples and hexane (1:10 *w*/*v*) were homogenized (6000 rpm, 1.5 min) using an IKA® T25 ultraturrax (IKA, Staufen im Breisgau, Germany), centrifuged (8500 rpm, 15 min, 20 °C) and supernatants collected. This procedure was repeated three times. Hexane was evaporated from pooled supernatants (25 rpm, 45 °C) and oil content was determined gravimetrically following the AOAC Official Method 960.39 [19]. Oil content was expressed as g per 100 g of freeze-dried kernels and cakes db (g/100 g db).

2.5.3. Microstructural Analysis

Control and PEF-treated kernels were fixed employing the procedure reported by Kendall et al. [21] with modifications. Glutaraldehyde and osmium tetroxide solutions were made using 0.1 M phosphate buffer (pH 7.2). Kernels were fixed in 2.5% glutaraldehyde solution and left overnight. Next, samples were washed three times in 0.1 M phosphate buffer and postfixed in 1.0% osmium tetroxide solution for 2 h. Subsequently, samples were washed twice with 0.1 M acetate buffer, incubated with 0.5% uranyl acetate for 30 min and rinsed two times in 0.1 M acetate buffer. Kernels were dehydrated in an acetonitrile series (30–100%) before embedding in epoxy EMbed 812 resin and polymerizing (48 h, 60 °C). A Reichert Jung Ultracut E microtome (Leica Microsystems, Washington, DC, USA) was used to obtain semithin and ultrathin sections. These sections were stained with Richardson blue [22] and examined at 20× and 100× by light microscopy in an Olympus BX41 microscope (Olympus, Allentown, PA, USA).

2.5.4. Lipoxygenase Activity

Kernels enzymatic extracts were obtained as described by Christopoulos and Tsantili [23]. An extraction solution was prepared by dissolving β-ME (5.0 mM), PVPP (1:100 *w/v*) and Triton X-114 (0.05:100 *w/v*) in 50 mM phosphate buffer (pH 6.6). Kernels (2.0 g) and the extraction solution (10 mL) were homogenized (6000 rpm, 40 s), filtered using glass wool and centrifuged (8000 rpm, 15 min, 4 °C). Supernatants were collected for determination of lipoxygenase (LOX) activity according to the procedure reported by Li et al. [24]. Solutions were made using 0.2 M borate buffer (pH 9.0). Linoleic acid dissolved in EtOH and 0.2 M borate buffer (1:1:1000 *v/v/v*) was employed as substrate stock solution to measure LOX activity. The stock solution (5 mL) was diluted completely in 20 mL of 0.2 M borate buffer and 5 mL of distilled water. The diluted solution (2 mL) and 0.2 M borate buffer (950 μL) were pipetted in a cell quartz and mixed by inversion. Next, enzymatic extracts (50 μL) were added and mixed by inversion. Absorbance was measured at 234 nm and registered every 10 s until 3 min of reaction in a Cecil CE 1010 UV-VIS spectrophotometer (Cecil Instruments Ltd., Cambridge, UK). LOX activity was calculated employing the molar extinction coefficient (ε) of hydroperoxides (26,800/M·cm) to express it as μmol of hydroperoxide produced per L of LOX per s (μmol/L·s) [25].

2.6. Oil Analysis

2.6.1. Acidity

Oil acidity was performed following the AOAC Official Method 940.28 [19] and results were expressed as mg KOH per 100 g of pecan nut oil (mg KOH/100 g).

2.6.2. Antioxidant Capacity

The DPPH radical scavenging capacity method reported by Gao et al. [26] was used to evaluate oil antioxidant capacity (AC). Pecan nut oil (1 mL) was diluted in ethyl acetate (10 mL) and mixed with 2 mL of 0.50 mM DPPH solution. The mixture was vortexed and left to react in darkness for 15 min. After this time, absorbance was measured at 515 nm. A standard curve of Trolox was prepared (0.003–0.030 mg/mL) to express results as mg Trolox equivalents per 100 g of pecan nut oil (mg Trolox EQ/100 g).

2.6.3. Oil Stability Index

The stability index (OSI) of extracted oils was measured in a Rancimat 679 apparatus (Metrohm AG, Herisau, Switzerland) [27]. Oil samples (3.0 g) were heated at 110 °C while air was bubbled at a flow rate of 20 L/h and volatile products were dissolved in deionized water (60 mL). OSI values were obtained by Metrodata Version 1.0 software (Metrohm AG, Switzerland) and expressed in hours (h).

2.6.4. Phytosterols Concentration

Extraction and quantification of phytosterols were done according to Domínguez-Avila et al. and Nair et al., respectively [28,29]. HCl and KOH solutions were made using EtOH as solvent. Oil was mixed with 6 M HCl (1:5 *w/v*) and incubated (1 h, 80 °C). The mixture was cooled in a water bath and 5 mL of 1.3% KOH was added and left to react (30 min, 80 °C). To phytosterols extraction, 2 mL of distilled water and 5 mL of hexane were included to the mixture, vortexed for 1 min, and centrifuged (3750 rpm, 15 min, 20 °C). The addition of distilled water and hexane was performed twice. Hexane was evaporated from pooled supernatants using a vacuum evaporator (2.5 h, 45 °C). Extracts were reconstituted in 0.5 mL of hexane for chromatographic analysis. A HPLC-ELSD system (Agilent 1200, Agilent Technologies, Santa Clara, CA, USA) equipped with a 5 μm, 4.6 mm × 500 mm Luna C8 column (Torrance, CA, USA) was employed to identify and quantify phytosterols from pecan nut oil. Column and ELSD temperature were maintained at 40 °C and 50 °C, respectively. Aliquots (10 μL) were analyzed employing a mobile phase consisted of MeOH:H_2O (95:5 v/v) at a flow rate of 1 mL/min and the detector was set at a gain of 16. Standard curves of β-sitosterol (0.2–1.2 mM), stigmasterol (0.2–1.2 mM), and campesterol (0.05–0.25 mM) were prepared to quantification. Concentrations were expressed as mg per kg of pecan nut oil (mg/kg).

2.7. Cakes Analysis

After the determination of cakes oil content, the defatted portion was used for phenolic compounds analysis. Total phenolics (TP), condensed tannins (CT), and AC of reference, control and PEF-treated cakes were performed following the procedures reported by Rábago-Panduro et al. [18]. Results were expressed as mmol equivalents per 100 g of defatted cake db: TP was expressed as mmol gallic acid EQ/100 g db, CT as mmol catechin EQ/100 g db and AC as mmol Trolox EQ/100 g db.

2.8. Statistical Analysis

PEF pretreatments, oil extraction and LOX activity were performed by triplicate while oil and cake analysis were done by duplicate to a total of six replicates per experimental group. One-way ANOVA and Dunnett test were performed with the Minitab® Version 18.1 software (Minitab, USA).

3. Results

3.1. Pecan Nut Kernels

3.1.1. Moisture and Oil Content

The moisture of dry pecan nuts used as reference was 2.4 ± 0.1 g/100 g db, while control kernels soaked at 3, 20 and 35 min had moisture contents of 13.5 ± 0.6, 21.1 ± 0.9, and 21.7 ± 0.7 g/100 g db, respectively. In PEF-treated kernels moisture contents also increased reaching values of 13.5 ± 1.0, 18.0 ± 0.1 and 19.3 ± 1.1 g/100 g db at 0.8, 7.8 and 15.0 kJ/kg pretreatments, respectively. However, the moisture of kernels processed at 7.8 and 15.0 kJ/kg was significantly lower than their respective control kernels (p = 0.004 and p = 0.015, respectively). Regarding the oil content, reference kernels contained 69.4 ± 0.7 g/100 g db, decreasing to 62.7 ± 0.0 g/100 g db in control kernels and 62.7 ± 0.7 g/100 g db in kernels pretreated by PEF, containing 9.7% less oil than reference kernels. Likewise, the o_{SW} was 5.7 ± 0.0 g for both control and PEF-treated kernels.

3.1.2. OEY, OEY_{TOTAL} and Microstructure

OEY and OEY_{TOTAL} of dry pecan nuts are displayed in Figure 1. The highest OEY was observed for reference kernels, being 88.7 ± 3.0%. Control kernels yielded 71.3 ± 1.0, 65.9 ± 3.1 and 72.3 ± 0.7% at 3, 20 and 35 min of soaking, respectively, while OEY of PEF-treated kernels was 74.3 ± 1.1, 69.1 ± 2.1 and 70.6 ± 3.2% at 0.8, 7.8 and 15.0 kJ/kg pretreatments, respectively. No statistical differences were observed between OEY of control and PEF-treated kernels submitted to comparable soaking times.

After considering the o_{SW}, OEY_{TOTAL} of control kernels increased up to 82.0 ± 1.0, 76.5 ± 3.1 and 83.0 ± 0.7% for 3, 20 and 35 min of soaking, respectively, and OEY_{TOTAL} of PEF-treated kernels rose to 85.0 ± 1.1, 79.8 ± 2.1 and 81.3 ± 3.2% for 0.8, 7.8 and 15.0 kJ/kg pretreatments, respectively (Figure 1). No statistical differences were observed between control and PEF-treated kernels. Kernels processed at 0.8 kJ/kg were selected for the microstructural analysis.

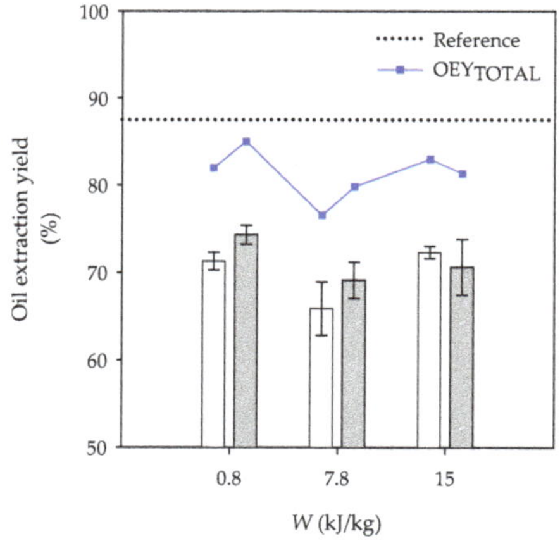

Figure 1. Oil extraction yields and total oil extraction yields (OEY_{TOTAL}) of dry pecan nuts. Reference kernels were not soaked nor PEF processed. Control kernels (□) were placed in tap water for 3, 20 and 35 min corresponding to PEF-treated kernels (■) processed at 0.8, 7.8 and 15.0 kJ/kg, respectively. OEY_{TOTAL} is the oil extraction yield estimated with the o_{SW}.

In order to compare the microstructure of PEF-treated and control kernels, a micrograph of the transversal section of dry kernels cotyledon tissue reported by Wakeling et al. was employed (Figure 2) [30]. In this micrograph, cells delimited by the cell wall containing intracellular oleosomes stained with toluidine blue were showed.

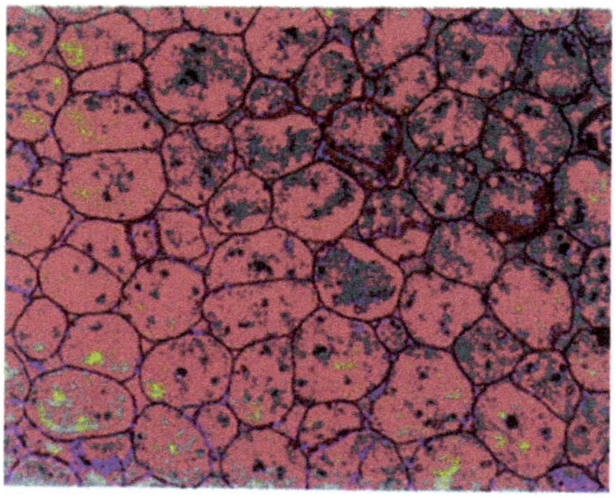

Figure 2. Light microscopy micrograph of cotyledon tissue of dry pecan nuts reported by Wakeling et al. [30] employed to compare the microstructure of control and PEF-treated kernels.

In the light microscopy of control kernels (Figure 3a), the cotyledon tissue exhibited similarities to the micrograph of Wakeling et al. [30] displaying delimited cells and oleosomes within the intracellular space stained with Richardson blue. However, at higher magnification, it appeared that the cotyledon tissue was composed of both intact and damaged cells. Additionally, oleosomes seemed to change their shape and aggregate within the intracellular space (Figure 3b). Concerning kernels pretreated by PEF, micrographs showed compaction of cells in testa and cotyledon tissues with a loss of delimited inclusions in the intracellular space (Figure 3c). It seems that PEF processing, rather than inducing the cell rupture, produced the rupture of intracellular inclusions. In Figure 3d, the higher magnification of PEF-treated kernels cotyledon tissue showed no difference between intact and damaged cells due to oleosomes fusion in the periphery of the cell.

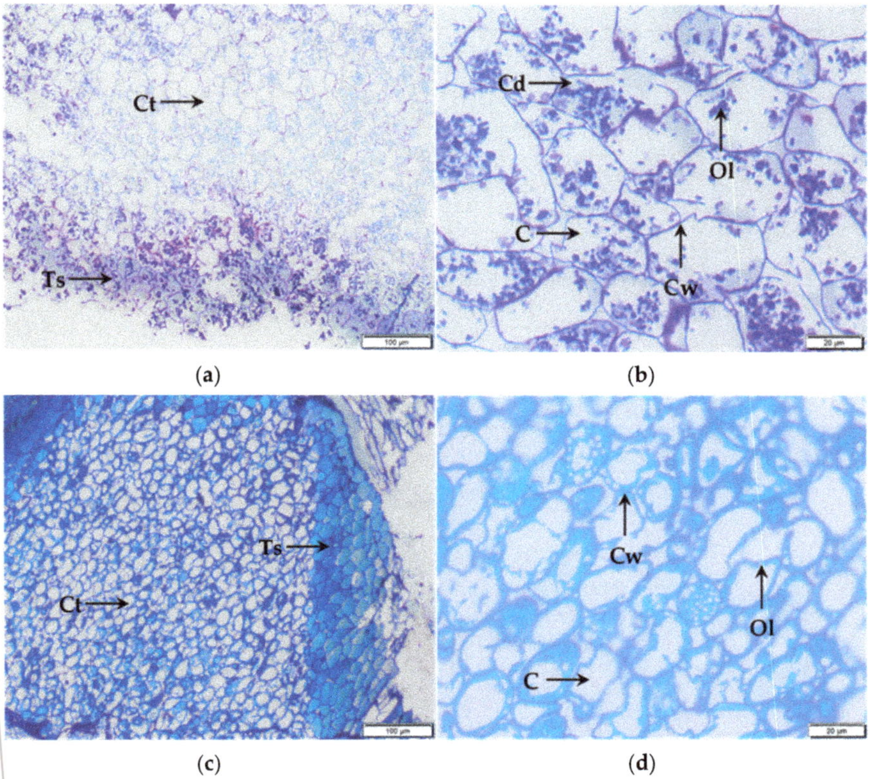

Figure 3. Light microscopy micrographs of testa and cotyledon tissue of dry pecan nuts. Control kernels (**a**,**b**) were placed in tap water for 3 min while PEF-treated kernels (**c**,**d**) were processed at 0.8 kJ/kg. Ts, testa; Ct, cotyledon tissue; C, cell; Cd, damaged cell; Cw, cell wall; Ol, oleosomes.

3.2. Pecan Nut Oil Stability

The acidity of oils extracted from dry pecan nuts ranged from 19.0 ± 0.9 to 21.8 ± 0.9 mg KOH/100 g (Table 1). Only the oil extracted from kernels processed at 15.0 kJ/kg was significantly different, being 14.2% higher than its control. Concerning AC of extracted oils, it varied from 45.7 ± 2.2 to 49.0 ± 1.7 mg Trolox/100 g with no significant differences between experimental groups (Table 1).

Based on acidity and AC results, OSI, phytosterols concentration and LOX activity were determined for kernels pretreated at 0.8 kJ/kg, their control kernels, and reference kernels. The OSI values were 10.4 ± 0.4, 10.9 ± 0.3 and 10.6 ± 0.5 h for reference, control and PEF-treated oils, respectively. Neither soaking nor PEF processing significantly changed the stability index of pecan nut oil. Regarding the phytosterols content, oil from reference, control and PEF-treated kernels contained a β-sitosterol concentration of 929.0 ± 89.3, 858.2 ± 62.6 and 910.5 ± 132.2 mg/kg, respectively, while stigmasterol ranged from 324.4 ± 48.5 to 501.5 ± 79.7 mg/kg (Table 2). Campesterol was not detected despite that it has been reported in low concentration in pecan nuts [31,32]. Neither β-Sitosterol nor stigmasterol significantly changed among extracted oils.

Table 1. Acidity and antioxidant capacity (AC) of oils extracted from dry pecan nuts.

		W (kJ/kg)					
	Reference	0.8		7.8		15.0	
		Control	PEF	Control	PEF	Control	PEF
Acidity mg KOH/100 g	21.4 ± 1.1	19.1 ± 0.9	19.9 ± 1.0	20.5 ± 0.1	21.8 ± 0.9	19.0 ± 0.9	21.7 ± 1.0 *
AC mg Trolox EQ/100 g	48.6 ± 0.8	45.7 ± 2.2	47.8 ± 1.3	49.0 ± 1.7	48.3 ± 2.8	48.2 ± 1.7	47.1 ± 1.2

W, specific energy input. Reference kernels were not soaked nor PEF processed. Control kernels were placed in tap water for 3, 20 and 35 min corresponding to PEF-treated kernels processed at 0.8, 7.8 and 15.0 kJ/kg, respectively. Oil acidity and AC were expressed per 100 g of pecan nut oil. Means with an asterisk indicated a significant difference from the control (α = 0.05).

Table 2. Phytosterols concentration of oils extracted from dry pecan nuts.

			W (kJ/kg)	
		Reference	0.8	
			Control	PEF
Phytosterols mg/kg	β-sitosterol	929.0 ± 89.3	858.2 ± 62.6	910.5 ± 132.2
	Stigmasterol	501.5 ± 79.7	352.1 ± 17.8	324.4 ± 48.5

W, specific energy input. Reference kernels were not soaked nor PEF processed. Control kernels were placed in tap water for 3 min corresponding to PEF-treated kernels processed at 0.8 kJ/kg. Concentrations were expressed per kg of pecan nut oil. Means were not significantly different from the control (α = 0.05).

The LOX activity of reference, control and PEF-treated kernels was 4.52 ± 0.14, 4.83 ± 0.10 and 4.41 ± 0.20 μmol/L·s, respectively. Reference and PEF-treated kernels showed LOX activity values significantly lower than control kernels ($p = 0.006$ and $p = 0.001$, respectively). It seemed that the application of PEF at 0.8 kJ/kg did not promote lipid oxidation of pecan nut oil. These results agreed with acidity, AC, and OSI values of the oil extracted from kernels processed at 0.8 kJ/kg. However, it should not be discarded lipid oxidation via enzyme activation since the acidity of pecan nut oil increased at higher W.

3.3. Pecan Nut Cakes

TP, CT and AC of cakes generated from dry pecan nuts are presented in Table 3. The reference cake showed TP, CT and AC values of 20.0 ± 1.0 mmol gallic acid EQ/100 g db, 15.4 ± 1.0 mmol catechin EQ/100 g db and 14.7 ± 1.0 mmol Trolox EQ/100 g db, respectively. TP varied from 18.1 ± 0.7 to 20.6 ± 1.8 mmol gallic acid EQ/100 g db in control cakes while, in PEF-treated cakes, TP decreased by 20.9% at 0.8 kJ/kg compared to its control (Table 3). The application of 0.8 and 7.8 kJ/kg increased the CT content of cakes by 27.0 and 10.7%, respectively, compared to their respective control cakes. An overall improvement in AC of cakes generated from PEF-treated kernels was observed with the highest increment observed at 0.8 kJ/kg, increasing by 24.3% in comparison with its control.

Table 3. Phenolic compounds and antioxidant capacity (AC) of cakes generated from dry pecan nuts.

		W (kJ/kg)					
	Reference	0.8		7.8		15.0	
		Control	PEF	Control	PEF	Control	PEF
Total phenolics mmol gallic acid EQ/100 g db	20.0 ± 1.0	20.6 ± 1.8	16.3 ±1.0 *	18.3 ± 2.1	17.9 ± 1.2	18.1 ± 0.7	16.5 ± 0.7 *
Condensed tannins mmol catechin EQ/100 g db	15.4 ± 1.0	17.4 ± 1.6	22.1 ± 2.1 *	13.1 ± 1.3	14.5 ± 0.8 *	13.3 ± 0.8	11.9 ± 1.0 *
AC mmol Trolox EQ/100 g db	14.7 ± 1.0	14.4 ± 1.7	17.9 ± 1.5 *	18.5 ± 1.2	19.2 ± 0.6	16.3 ± 0.4	18.1 ± 1.3 *

W, specific energy input. Reference kernels were not soaked nor PEF processed. Control kernels were placed in tap water for 3, 20 and 35 min corresponding to PEF-treated kernels processed at 0.8, 7.8 and 15.0 kJ/kg, respectively. Total phenolics, condensed tannins, and AC were expressed per 100 g of defatted cake. Means with an asterisk indicated a significant difference from the control (α = 0.05).

4. Discussion

Dry pecan nuts used as reference kernels displayed moisture and oil content within ranges reported in other studies [10]. Regarding kernels' immersion in water, moisture reductions observed in kernels pretreated at 7.8 and 15.0 kJ/kg could be associated with the release of intracellular water as a result of electroporation of the cell membrane during PEF processing [33,34]. Oil content reductions along with o_{SW} values were comparable to those reported for fresh kernels pretreated by PEF [18], being related to the exposure of kernels cotyledon tissue to the soaking water provoking oleosomes release. An increment of oleosomes extraction was expected for dry kernels compared to the fresh ones as a result of the drying effect on kernels' microstructure [11]. However, given that the o_{SW} was comparable among pecan nuts, it is proposed that kernels rehydration and water-soluble compounds release were favored over oleosomes extraction.

Concerning oil extraction from dry pecan nuts, the OEY of reference kernels (88.7%) was higher than those reported for pecan nut oil extracted by hydraulic pressing (56.4–58.9%) [9,35,36] and higher than the OEY of fresh pecan nut kernels used as a reference in the previous study (63.8%) [18]. Dry kernels differed from fresh kernels in moisture and oil content but also in kernels microstructure displaying more structural damage compared to fresh pecan nuts [37–39]. Savoire et al. evaluated oil extraction processes comparing seeds from the same type that only differed in moisture and oil content. The authors reported an overall improvement of OEY in seeds with lower moisture and higher oil content [40]. They also stated that varietal differences that influence seed characteristics (moisture and oil content, hull and testa thinness, and pore number and size) could affect oil extraction by modifying seed pressing behavior, oil flow and kernels permeability.

The PEF pretreatments did not increase oil extraction from dry pecan nut kernels, contrasting with previous findings where the application of 0.8 kJ/kg increased OEY_{TOTAL} of fresh kernels up to 74.8% being higher than OEY of kernels without soaking nor PEF pretreatment [18]. Sarkis et al. reported similar results where drying and grinding of sesame seeds yielded higher values than the seeds pretreated by PEF [4]. Nikiforidis stated that soaking of seeds causes cell swelling, changing oleosomes shape and diffusion kinetics [12]. Consequently, the soaking of dry pecan nuts might hinder oil extraction by the initial reduction of oil content caused by oleosomes release and the reorganization of the remaining oleosomes within the cotyledon tissue trapping them within kernels microstructure.

The analysis of oil stability showed that oils extracted from dry pecan nuts were within the acidity range accepted by the Food and Agriculture Organization for cold-pressed oils [41]. Furthermore, the AC of extracted oils was slightly lower than those achieved in the study employing fresh kernels but higher than those reported for pecan nut oil obtained by solvent extraction [18,28]. Regarding OSI values, Oro et al. reported a similar OSI value for pecan nut oil also extracted by mechanical pressing [42]. Phytosterols concentration of oil extracted from reference kernels was lower in comparison with other studies where pecan nut oil was characterized [8,43,44]. Differences in acidity, AC, OSI and phytosterols concentration might be related to varietal differences between pecan nuts as well as tocopherols concentration since these compounds have been associated to the antioxidant capacity of pecan nut oil [40,44]. Concerning the LOX activity of reference kernels, this value was similar to those reported for different cultivars of walnuts [45]. In tree nuts, LOX inactivation has been associated to moisture reduction [11,46]. Thus, it is possible that the increment of LOX activity observed in control kernels could be related to moisture increase by modifying cell and organelle structure. Instead, LOX activity maintenance observed in PEF-treated kernels compared to control kernels might be due to the improvement of condensed tannins which have been reported to inhibit LOX [47–49].

Total phenolics, condensed tannins, and antioxidant capacity of cakes generated from dry kernels were lower than those reported for fresh kernels [18]. These differences could be due to the effect of drying as this process has been demonstrated to decrease TP, CT and AC of pecan nuts [50]. Despite that, the application of PEF increased not only the

CT concentration but also enhanced AC of the cakes. Considering that pecan nuts AC is closely related to their CT concentration, the increment of AC in cakes from PEF-treated kernels might be attributed to the release of simple phenolic compounds while condensed tannins are retained in the cake increasing its AC [50,51].

5. Conclusions

The application of PEF to dry pecan nuts did not increase OEY nor OEY_{TOTAL} of pretreated kernels (69.1–74.3% and 79.8–85.0%, respectively) in comparison with kernels without soaking nor PEF processing (88.7%). These results could be due to changes in oleosomes characteristics and their localization within the cotyledon tissue of dry kernels submitted to water immersion. Kernels processed at 0.8 kJ/kg were used for the microstructural analysis showing a reduction of the cell size and oleosomes fusion in the intracellular space. These changes might help oil flow during mechanical extraction. Comparable values of acidity, AC, OSI and LOX activity between extracted oils showed that the application of PEF did not negatively affect pecan nut oil quality. Furthermore, the 0.8 kJ/kg pretreatment increased CT and AC of generated cakes compared to cakes from untreated kernels and kernels soaked for 3 min. These results showed that the effectiveness of PEF to increase OEY_{TOTAL} of pecan nuts is dependent on not only kernels' moisture and oil content but also on the drying process. It appears that the rehydration of dry kernels produced a negative effect on their microstructure and oleosomes characteristics. Thus, further research focused on these variables is necessary along with other microscopy techniques and analytical procedures to better understand the effect of PEF pretreatments on oil extraction from pecan nuts.

Author Contributions: Conceptualization, L.M.R.-P. and M.M.-d.l.P.; resources, M.P.R.-F., O.M.-B. and J.W.-C., data curation, L.M.R.-P. and J.W.-C.; writing—original draft preparation, L.M.R.-P.; Writing—review and editing, M.M.-d.l.P., M.P.R.-F., O.M.-B. and J.W.-C. All authors have read and agreed to the published version of the manuscript.

Funding: This research was funded by Tecnológico de Monterrey and Consejo Nacional de Ciencia y Tecnología (CONACyT) scholarship programs (CVU 418204).

Data Availability Statement: The authors confirm that the data supporting the findings of this study are available within the article.

Conflicts of Interest: The authors declare no conflict of interest.

References

1. Puértolas, E.; Koubaa, M.; Barba, F.J. An overview of the impact of electrotechnologies for the recovery of oil and high-value compounds from vegetable oil industry: Energy and economic cost implications. *Food Res. Int.* **2016**, *80*, 19–26. [CrossRef]
2. Guderjan, M.; Elez-Martínez, P.; Knorr, D. Application of pulsed electric fields at oil yield and content of functional food ingredients at the production of rapeseed oil. *Innov. Food Sci. Emerg. Technol.* **2007**, *8*, 55–62. [CrossRef]
3. Moradi, N.; Rahimi, M. Effect of simultaneous ultrasound/pulsed electric field pretreatments on the oil extraction from sunflower seeds. *Sep. Sci. Technol.* **2018**, *53*, 2088–2099. [CrossRef]
4. Sarkis, J.R.; Boussetta, N.; Tessaro, I.C.; Marczak, L.D.F.; Vorobiev, E. Application of pulsed electric fields and high voltage electrical discharges for oil extraction from sesame seeds. *J. Food Eng.* **2015**, *153*, 20–27. [CrossRef]
5. Veneziani, G.; Esposto, S.; Taticchi, A.; Selvaggini, R.; Sordini, B.; Lorefice, A.; Daidone, L.; Pagano, M.; Tomasone, R.; Servili, M. Extra-virgin olive oil extracted using pulsed electric field technology: Cultivar impact on oil yield and quality. *Front. Nutr.* **2019**, *6*, 1–8. [CrossRef] [PubMed]
6. Alvarez-Parrilla, E.; Urrea-López, R.; de la Rosa, L.A. Bioactive components and health effects of pecan nuts and their by-products: A review. *J. Food Bioact.* **2018**, *1*, 56–92. [CrossRef]
7. Atanasov, A.G.; Sabharanjak, S.M.; Zengin, G.; Mollica, A.; Szostak, A.; Simirgiotis, M.; Huminiecki, Ł.; Horbanczuk, O.K.; Nabavi, S.M.; Mocan, A. Pecan nuts: A review of reported bioactivities and health effects. *Trends Food Sci. Technol.* **2018**, *71*, 246–257. [CrossRef]
8. Fernandes, G.D.; Gómez-Coca, R.B.; Pérez-Camino, M.d.C.; Moreda, W.; Barrera-Arellano, D. Chemical characterization of major and minor compounds of nut oils: Almond, hazelnut, and pecan nut. *J. Chem.* **2017**, *2017*, 1–11. [CrossRef]
9. Scapinello, J.; Magro, J.D.; Block, J.M.; Di Luccio, M.; Tres, M.V.; Oliveira, J.V. Fatty acid profile of pecan nut oils obtained from pressurized n-butane and cold pressing compared with commercial oils. *J. Food Sci. Technol.* **2017**, *54*, 3366–3369. [CrossRef]

10. Alasalvar, C.; Shahidi, F. *Tree Nuts: Composition, Phytochemicals, and Health Effects*; CRC Press: Boca Raton, FL, USA, 2008; ISBN 9780849337352.
11. Perren, R.; Escher, F.E. Impact of roasting on nut quality. In *Improving the Safety and Quality of Nuts*; Elsevier: Amsterdam, The Netherlands, 2013; pp. 173–197, ISBN 9780857092663.
12. Nikiforidis, C.V. Structure and functions of oleosomes (oil bodies). *Adv. Colloid Interface Sci.* **2019**, *274*, 1–6. [CrossRef]
13. Buranasompob, A.; Tang, J.; Powers, J.R.; Reyes, J.; Clark, S.; Swanson, B.G. Lipoxygenase activity in walnuts and almonds. *LWT Food Sci. Technol.* **2007**, *40*, 893–899. [CrossRef]
14. Heldt, H.-W.; Piechulla, B.; Heldt, F. *Plant Biochemistry*, 4th ed.; Elsevier: Amsterdam, The Netherlands, 2011; ISBN 9780123849861.
15. Shahidi, F.; Zhong, H.J. Methods for measuring lipid oxidation. In *Bailey's Industrial Oil and Fat Products*; Wiley: Hoboken, NJ, USA, 2020; pp. 1–27. ISBN 047167849X.
16. Shahidi, F.; John, J.A. *Oxidation and Protection of Nuts and Nut Oils*; Woodhead Publishing Limited: Sawston, UK, 2010; Volume 1.
17. Subroto, E.; Manurung, R.; Heeres, H.J.; Broekhuis, A.A. Optimization of mechanical oil extraction from *Jatropha curcas* L. kernel using response surface method. *Ind. Crops Prod.* **2015**, *63*, 294–302. [CrossRef]
18. Rábago-Panduro, L.M.; Morales-de la Peña, M.; Martín-Belloso, O.; Welti-Chanes, J. Application of pulsed electric fields (PEF) on pecan nuts [*Carya illinoinensis* (Wangenh. K. Koch)]: Oil extraction yield and compositional characteristics of the oil and its by-product. *Food Eng. Rev.* **2021**. [CrossRef]
19. AOAC International. *Official Methods of Analysis*; AOAC International: Rockville, MA, USA, 1996.
20. Villarreal-Lozoya, J.E.; Lombardini, L.; Cisneros-Zevallos, L. Phytochemical constituents and antioxidant capacity of different pecan [*Carya illinoinensis* (Wangenh.) K. Koch] cultivars. *Food Chem.* **2007**, *102*, 1241–1249. [CrossRef]
21. Kendall, C.W.; Jenkins, D.J.; Parker, C.; Ellis, P.R.; Pacy, J.F.; Ren, Y.; Waldron, K.W. Role of cell walls in the bioaccessibility of lipids in almond seeds. *Am. J. Clin. Nutr.* **2018**, *80*, 604–613. [CrossRef]
22. Richardson, K.C.; Jarett, L.; Finke, E.H. Embedding in epoxy resins for ultrathin sectioning in electron microscopy. *Stain. Technol.* **1960**, *35*, 313–323. [CrossRef] [PubMed]
23. Christopoulos, M.V.; Tsantili, E. Participation of phenylalanine ammonia-lyase (PAL) in increased phenolic compounds in fresh cold stressed walnut (*Juglans regia* L.) kernels. *Postharvest Biol. Technol.* **2015**, *104*, 17–25. [CrossRef]
24. Li, Y.-Q.; Tian, W.-L.; Mo, H.-Z.; Zhang, Y.-L.; Zhao, X.-Z. Effects of pulsed electric field processing on quality characteristics and microbial inactivation of soymilk. *Food Bioprocess Technol.* **2012**, *6*, 1907–1916. [CrossRef]
25. Gardner, H.W. Analysis of lipoxygenase activity and products. *Curr. Protoc. Food Anal. Chem.* **2001**. [CrossRef]
26. Gao, P.; Liu, R.; Jin, Q.; Wang, X. Comparative study of chemical compositions and antioxidant capacities of oils obtained from two species of walnut: *Juglans regia* and *Juglans sigillata*. *Food Chem.* **2019**, *279*, 279–287. [CrossRef]
27. Tovar, M.J.; Romero, M.P.; Alegre, S.; Girona, J.; Motilva, M.J. Composition and organoleptic characteristics of oil from Arbequina olive (*Olea europaea* L.) trees under deficit irrigation. *J. Sci. Food Agric.* **2002**, *82*, 1755–1763. [CrossRef]
28. Domínguez-Avila, J.A.; Alvarez-Parrilla, E.; González-Aguilar, G.A.; Villa-Rodríguez, J.; Olivas-Orozco, G.I.; Molina Corral, J.; Gómez-García, M.D.C.; de la Rosa, L.A. Influence of growing location on the phytochemical content of pecan (*Carya illinoinensis*) oil. *J. Food Res.* **2013**, *2*, 143–152. [CrossRef]
29. Nair, V.D.P.; Kanfer, I.; Hoogmartens, J. Determination of stigmasterol, β-sitosterol, and stigmastanol in oral dosage forms using high performance liquid chromatography with evaporative light scattering detection. *J. Pharm. Biomed. Anal.* **2006**, *41*, 731–737. [CrossRef] [PubMed]
30. Wakeling, L.T.; Mason, R.L.; D'Arcy, B.R.; Caffin, N.A.; Gowanlock, D. Microscopic structure of opalescent and nonopalescent pecans. *J. Food Sci.* **2003**, *68*, 2238–2242. [CrossRef]
31. Phillips, K.M.; Ruggio, D.M.; Ashraf-Khorassani, M. Phytosterol composition of nuts and seeds commonly consumed in the United States. *J. Agric. Food Chem.* **2005**, *53*, 9436–9445. [CrossRef]
32. U.S. Department of Agriculture FoodData Central. Available online: https://fdc.nal.usda.gov/ (accessed on 31 March 2020).
33. Ribas-Agustí, A.; Martín-Belloso, O.; Soliva-Fortuny, R.; Elez-Martínez, P. Enhancing hydroxycinnamic acids and flavan-3-ol contents by pulsed electric fields without affecting quality attributes of apple. *Food Res. Int.* **2019**, *121*, 433–440. [CrossRef]
34. Tylewicz, U.; Aganovic, K.; Vannini, M.; Toepfl, S.; Bortolotti, V.; Dalla Rosa, M.; Oey, I.; Heinz, V. Effect of pulsed electric field treatment on water distribution of freeze-dried apple tissue evaluated with DSC and TD-NMR techniques. *Innov. Food Sci. Emerg. Technol.* **2016**, *37*, 352–358. [CrossRef]
35. Costa-Singh, T.; Jorge, N. Characterization of *Carya illinoiensis* and *Juglans regia* oils obtained by different extraction systems. *Acta Sci. Technol.* **2015**, *37*, 279. [CrossRef]
36. Polmann, G.; Badia, V.; Frena, M.; Teixeira, G.L.; Rigo, E.; Block, J.M.; Camino Feltes, M.M. Enzyme-assisted aqueous extraction combined with experimental designs allow the obtaining of a high-quality and yield pecan nut oil. *LWT Food Sci. Technol.* **2019**, *113*, 1–7. [CrossRef]
37. Gutiérrez, L.-F.; Ratti, C.; Belkacemi, K. Effects of drying method on the extraction yields and quality of oils from quebec sea buckthorn (*Hippophaë rhamnoides* L.) seeds and pulp. *Food Chem.* **2008**, *106*, 896–904. [CrossRef]
38. Jia, X.; Luo, H.; Xu, M.; Wang, G.; Xuan, J.; Guo, Z. Investigation of nut qualities of pecan cultivars grown in China. *J. Plant Sci.* **2019**, *7*, 117–124. [CrossRef]
39. Santerre, C.R. *Pecan Technology*; Champman & Hall: London, UK, 1994; ISBN 9789401095945.

40. Savoire, R.; Lanoisellé, J.L.; Vorobiev, E. Mechanical continuous oil expression from oilseeds: A review. *Food Bioprocess Technol.* **2013**, *6*, 1–16. [CrossRef]
41. Food and Agriculture Organization of the United Nations SECTION 2. Codex Standards for Fats and Oils from Vegetable Sources. Available online: http://www.fao.org/3/y2774e/y2774e04.htm (accessed on 29 January 2020).
42. Oro, T.; Bolini, H.M.A.; Barrera-Arellano, D.; Block, J.M. Physicochemical and sensory quality of crude brazilian pecan nut oil during storage. *JAOCS J. Am. Oil Chem. Soc.* **2009**, *86*, 971–976. [CrossRef]
43. Al Juhaimi, F.; Özcan, M.M.; Ghafoor, K.; Babiker, E.E.; Hussain, S. Comparison of cold-pressing and soxhlet extraction systems for bioactive compounds, antioxidant properties, polyphenols, fatty acids and tocopherols in eight nut oils. *J. Food Sci. Technol.* **2018**, *55*, 3163–3173. [CrossRef] [PubMed]
44. Gong, Y.; Pegg, R.B.; Carr, E.C.; Parrish, D.R.; Kellett, M.E.; Kerrihard, A.L. Chemical and nutritive characteristics of tree nut oils available in the U.S. market. *Eur. J. Lipid Sci. Technol.* **2017**, *119*, 1–15. [CrossRef]
45. Yuan, X.; Huang, S.; Ma, H.; Huang, N.; Ye, N. Differential responses of walnut cultivars to cold storage and their correlation with postharvest physiological parameters. *Hortic. Environ. Biotechnol.* **2019**, *60*, 345–356. [CrossRef]
46. Uquiche, E.; Jeréz, M.; Ortíz, J. Effect of pretreatment with microwaves on mechanical extraction yield and quality of vegetable oil from Chilean hazelnuts (*Gevuina avellana* Mol). *Innov. Food Sci. Emerg. Technol.* **2008**, *9*, 495–500. [CrossRef]
47. do Prado, A.C.P.; Manion, B.A.; Seetharaman, K.; Deschamps, F.C.; Barrera Arellano, D.; Block, J.M. Relationship between antioxidant properties and chemical composition of the oil and the shell of pecan nuts [*Carya illinoinensis* (Wangenh) C. Koch]. *Ind. Crops Prod.* **2013**, *45*, 64–73. [CrossRef]
48. Shahidi, F.; Ambigaipalan, P. Phenolics and polyphenolics in foods, beverages and spices: Antioxidant activity and health effects—A review. *J. Funct. Foods* **2015**, *18*, 820–897. [CrossRef]
49. Bouali, I.; Tsafouros, A.; Ntanos, E.; Albouchi, A.; Boukhchina, S.; Roussos, P.A. Inter-cultivar and temporal variation of phenolic compounds, antioxidant activity and carbohydrate composition of pecan (*Carya illlinoinensis*) kernels grown in Tunisia. *Hortic. Environ. Biotechnol.* **2020**, *61*, 183–196. [CrossRef]
50. Rábago-Panduro, L.M.; Martín-Belloso, O.; Welti-Chanes, J.; Morales-de la Peña, M. Changes in bioactive compounds content and antioxidant capacity of pecan nuts [*Carya illinoinensis* (Wangenh. K. Koch)] during storage. *Rev. Mex. Ing. Química* **2020**, *19*, 1439–1452. [CrossRef]
51. Flores-Martínez, H.; León-Campos, C.; Estarrón-Espinosa, M.; Orozco-Ávila, I. Process optimization for the extraction of antioxidants from mexican oregano (*Lippia graveolens* HBK) by the response surface methodology (RSM) approach. *Rev. Mex. Ing. Química* **2016**, *15*, 773–785. [CrossRef]

foods

MDPI

Article

Partial Characterization of the Impact of Saffron on the Sensory and Physicochemical Quality Traits of Dry-Cured Ham

Elena M. Gómez-Sáez [1], Natalia Moratalla-López [2], Gonzalo L. Alonso [2] and Herminia Vergara [3,*]

1 Benibaldo, S.A.U., 02007 Albacete, Spain; Calidad@benibaldo.com
2 Cátedra de Química Agrícola, ETSI Agrónomos y de Montes, Campus Universitario, Universidad de Castilla-La Mancha, 02006 Albacete, Spain; Natalia.Moratalla@uclm.es (N.M.-L.); Gonzalo.Alonso@uclm.es (G.L.A.)
3 Science and Agroforestry Technology and Genetic Department, Higher Technical School of Agricultural Engineering and Forestry, Food Quality Section-Regional Development Institute, University of Castilla-La Mancha, 02006 Albacete, Spain
* Correspondence: Herminia.Vergara@uclm.es; Tel.: +34-681361088

Abstract: This study determined the effect of adding three concentrations of saffron (A: high, B: medium, and C: low) on vacuum-packaged dry-cured ham slices. The pH and the color coordinates were assessed at 0, 7, 14, 28 and 60 days of storage, and sensorial quality (visual appearance, odor and flavor) and safranal content were analyzed at 7, 14, 28 and 60 days. Saffron concentration did not significantly affect the pH or color (except in a* (redness) and b* (yellowness) at day 28; $p < 0.001$). Storage period affected pH values ($p < 0.001$) in all groups with a significant decline from day 28 ($p < 0.05$); the color coordinates showed a high stability (only L* (lightness) varied in the C group samples; $p < 0.01$). Sensorial quality did not vary with the time in any group. Significant differences were found among groups in visual appearance ($p < 0.05$) and flavor ($p < 0.001$) at day 14 and in odor at day 14, 28, and 60. In general, the C group samples obtained the highest scores. Safranal content varied significantly with the time in a different way in each group, with differences among groups at day 14 and 60 ($p < 0.001$).

Keywords: ham; slices; *Crocus sativus* L.; pH; color; sensorial quality; safranal

Citation: Gómez-Sáez, E.M.; Moratalla-López, N.; Alonso, G.L.; Vergara, H. Partial Characterization of the Impact of Saffron on the Sensory and Physicochemical Quality Traits of Dry-Cured Ham. *Foods* **2021**, *10*, 1506. https://doi.org/10.3390/foods10071506

Academic Editor: Mohammed Gagaoua

Received: 24 May 2021
Accepted: 25 June 2021
Published: 29 June 2021

1. Introduction

Spain ranks second in the European Union as regards pork production (4530,480–24,075,087 t) [1]. Among Spanish pork products, the most popular is the dry-cured ham (Jamón), the consumption per capita of which exceeds 1.60 kg [2]; it is typically offered as cured ham slices sold in trays owing to the increasing consumer demand for ready-to-eat products [3]. Dry-cured ham is a meat product highly appreciated by consumers because of its sensory characteristics made with pig hind limbs processed under traditional practices [4] that include salting, washing, draining, drying, and curing. This meat product is available under four official labels: "Jamón Serrano Traditional Specialty Guaranteed (TSG)", "Jamón de Trévelez Protected Geographical Indication (PGI)", "Jamón de Serón Protected Geographical Indication" and "Jamón de Teruel Protected Designation of Origin (PDO)".

Meat product quality is determined by physicochemical, sensory, and hygienic-sanitary properties [5], and many factors can affect these parameters in dry-cured ham, such as raw material [6] or processing technologies [7] such as salting [8] cutting [9], and drying, which has an effect on texture [10].

Seasoning is used to aromatize meat products and make them safe from a microbiological and physicochemical perspective [11,12]. Unlike other Spanish meat products [13,14], which are manufactured with the most popular spices (white and black pepper, garlic, and paprika), dry-cured ham is typically seasoned only with salt. Other additives are sometimes used, such as sugar, antioxidants (E-301), preservatives (E-250 and E-252), and acidity corrector (E-331iii). However, saffron (the dried stigmas of *Crocus sativus* L.), one of the most important flavoring spices in Spain, has not yet been used. Some studies confirm that saffron alleviates inflammatory diseases such as diabetes [15] and cardiovascular diseases [16] and has preventive effects on cancer [17,18]. Saffron is composed of a group of carotenoids, crocetin sugar esters, picrocrocin, and a wide array of ketones and terpenic aldehydes, with safranal being the most important compound [19–23] that contributes to more than 70% of the aroma of Spanish saffron [22]. Safranal (2,6,6-trimethyl-1,3-cyclohexadiene-1-carboxaldehyde), which results from the hydrolysis of picrocrocin [24], is credited with specific bioactive effects, such as satiety-inducing, antidepressant, and neuroprotective effects [19,25] and protective effects on ischemia-induced PC12 cell injury through inhibition of oxidative stress and apoptosis [26]; safranal may also be used in future research on the treatment of schizophrenia [27].

Currently, saffron, which its use dates back to the Sumerians [28], is added to the main food dishes in different Mediterranean countries [28] as a natural food additive for coloring and flavoring [29], without limitation in culinary purposes (Regulation (EC) No 1333/2008 of the European Parliament and of The Council of 16 December 2008 on Food Additives). Saffron has no toxic effects when is used in culinary quantities [30].

To date, there are no reports on the use of saffron to seasone meat products such as dry-cured ham, however, saffron has been used to flavor cheese [31,32] and yoghurt [33]. When saffron is used to enhance the flavor of foods, it is used in very small concentrations so as not to detract from the flavor of the main product. Therefore, this study was carried out to investigate the effect of adding low concentrations of saffron by impregnation of sliced cured ham on the sensorial acceptance and physicochemical quality during the storage period. In addition, the transfer of aromatics from saffron to the product was assessed by analyzing the safranal content using headspace-stir bar sorptive extraction–gas chromatography/mass spectrometry (HS-SBSE–GC-MS).

The results of this study will contribute to the meat industry through the discovery of innovative products that may provide added value and have favorable health effects on consumers.

2. Materials and Methods

2.1. Experimental Design

In this study, 10 dry-cured hams (8 ± 1 kg and pH > 5.6)—from 5 Duroc female pigs—belonging to the official label "Jamón Serrano TSG" were used. The pigs were raised under intensive conditions and in compliance with animal welfare standard [34]. Transportation of hams from the slaughterhouse and cutting rooms to the manufacturing industry (provider of Benibaldo S.A.U., Albacete, Spain) was conducted in refrigerated vehicles at a temperature <3 °C. Then, the dry-cured hams were processed using the following protocol: hams were pitted, peeled, polished, knocked out, and sliced in a slicer (Model USA-280, José Bernad, S.L., Albacete, Spain). The slices (0.8 ± 0.1 mm thickness) were placed on a coating base until 100 g was reached.

Because there are no previous studies on the addition of saffron to meat products, to establish the concentrations of this spice in each group, first, a preliminary sensory analysis was performed using a triangle test, to understand whether panelists can differentiate between the visual appearance of samples spiced with the lowest saffron concentration (0.015% *w*/*w*) and samples without saffron (control group). A sensory analysis was conducted following the recommendations made in a previous study [35], and the results were statistically analyzed according to [36]. According to a previous study in which 30 panelists participated in such a sensory analysis, the minimum number of correct answers for determining a perceptible difference should be 19 (α = 0.1%). In the present study, 28 of 30 panelists answered correctly. Thus, this concentration of saffron was considered the lowest that should be added to the ham slices. Therefore, the following groups were compared:

A (high: 0.055% w/w), B (medium: 0.035% w/w) and C (low: 0.015% w/w) and a control group without saffron.

Ground saffron, under the PDO label "Azafrán de La Mancha" was directly purchased from a producer (Agrícola Técnica de Manipulación y Comercialización, Minaya, Albacete, Spain). Generally, this product is commercially available in stigma form and not in powder form. Ground saffron was characterized according to ISO 3632:2011 [37] ($A^{1\%}_{1\ cm}$ 440 nm = 230 ± 2, $A^{1\%}_{1\ cm}$ 257 nm = 95 ± 3, and $A^{1\%}_{1\ cm}$ 330 nm = 24 ± 1). Saffron was evenly added to the samples using a stainless-steel dredger (Model KCFINE, Kitchen Craft, 7.3 × 7.3 × 9.1, 140 g, Amazon, Spain). The temperature during the manufacturing process did not exceed 15 °C. Samples (sachets of ham slices of 100 g each) were packed under vacuum conditions with a packaging machine (Model JB-350/M, José Bernad, S.L.) using a base to plate ham (Model 16409, 26 cm, Manchaplas, S.L., Albacete, Spain) and vacuum bags (Model 90M, 350 × 300 mm^2, Gutplask, S.L., Getafe, Madrid, Spain) with an oxygen permeability rate <70 $cm^3/m^2/24$ h, tensile strength at break of 21–43 MPa, elongation at break of 400–600%, and a slow resistance to penetration >1 N. After packaging, samples were stored in the dark at 2 °C until the analysis. Physicochemical quality was analyzed at 0, 7, 14, 28, and 60 days of storage, whereas the sensory analysis was done after 7 days of preparation. A total of 192 sachets were prepared, of which 20 and 160 were used in the physicochemical and sensory analysis, respectively, and 12 were used to analyze the transfer of aromatics.

2.2. *Analysis of Samples*

2.2.1. Physicochemical Quality (pH and Color Parameters)

To determine pH values, a pH meter (Crison GLP 22 + pH & Ion-meter-Crison Instruments, S.A., Barcelona, Spain) connected to a penetration electrode was used. pH was directly measured on five different slices randomly selected from each sachet.

Color coordinates (L*, lightness; a*, redness; and b*, yellowness) were evaluated using a CR 400 chroma meter (Minolta, Osaka, Japan) with a D65 illuminant and 10° standard observer, calibrated against a standard white tile. In all ham groups, five measurements were randomly taken on the surface of the sample on each sachet, and the mean value of three measurements was used. Chroma [$C^* = (a^2 + b^2)^{1/2}$] and hue angle ($h^* = \tan^{-1}$ $(b^*/a^*)^\circ$ were calculated [38].

2.2.2. Sensorial Quality

To measure the degree of acceptance or rejection of the three groups of flavored ham, a hedonic test was performed at 7, 14, 28, and 60 days of storage, at mid-morning in the test room of the university for 45 minutes approximately. It was carried out by 30 panelists (the same ones who participated in the triangular test described above; regular consumers of dry-cured ham; between 20 and 70 years old, 48% women, belonging to the university community). The attributes to evaluate were: Visual appearance: color assessment relating to the red color and presence of saffron. Odor: assessment relating to the characteristic odor associated with curing process and mixed with saffron. Flavor: assessment relating to the characteristic taste associated with the salt and curing process mixed with saffron. Samples were kept at environmental temperature for half an hour before the tasting. Three flavored dry-cured ham slices, one from each group, were placed in plastic plates and codified with three random numbers. Cold water and toasted bread were supplied to each panelist before testing each sample for cleansing the palate. Panelists, untrained consumers, were instructed at the beginning of each session for 15 minutes. The test they were to perform and how to proceed after eating each slice of flavored ham was explained to them.

The samples were rated on a 5-point hedonic scale, as follow: 1 = "Do not like it", 2 = "Slightly dislike it", 3 = "Neither like it nor dislike", 4 = "Like it" and 5 denoted "I like it very much". The consumers chose the expression in relation to their perception and acceptance of the flavored group. Then, the panelists indicated the concentration they liked the most overall.

2.2.3. Analysis of Safranal in Dry-Cured Ham

The transfer of aromatics from saffron to the meat product—flavored dry-cured ham with this spice—was analyzed by HS-SBSE–GC-MS. The volatile compounds were desorbed from a polydimethylsiloxane-coated stir bar (0.5 mm film thickness × 20 mm length; Twister, Gerstel GmbH (Mülheim an der Ruhr, Germany) using an automated thermal desorption unit (TDU, Gerstel) mounted on an Agilent 7890A gas chromatography system coupled to a quadrupole Agilent 5975C electron ionization mass spectrometric detector (Agilent Technologies, Palo Alto, CA, USA) equipped with a fused silica capillary column (BP21 stationary phase; 30 m length, 0.22 mm internal diameter, and 0.25 μm film thickness; SGE, Ringwood, Australia). The carrier gas was helium with a constant column pressure of 20.75 psi. From each group, 200 mg of flavored dry-cured ham was used (every sachet was divided into four equal parts and 25 mg from each part was used) for each time point (7, 14, 28, and 60 days of storage). These 200 mg were analyzed in triplicate to detect and quantify the major component of saffron (safranal), which is the main compound that can be used to distinguish and classify cured ham flavored with saffron [22]. Thus, 36 vials of 10 mL were used, and the method validated in a previous study [22] was used to analyze the transfer of aromatics from saffron to dry-cured ham.

Mass spectrometry data acquisition was performed in the positive scan mode; however, to avoid matrix interferences, the MS quantification was performed in the SIM mode using the major ion of safranal.

2.3. Statistical Analysis

Data were analyzed using the statistical package SPSS 24.0 version (SPSS Inc., Chicago, IL, USA). To analyze the effect of saffron concentration (A: high, B: medium, and C: low) on the physicochemical parameters (pH and color), sensorial quality (visual appearance, odor, and flavor), and safranal transfer, a Shapiro-Wilk test was carried out to check the normality and a Levene's test of homogeneity of variance of all values, then, a one-way analysis of variance (ANOVA) was performed. Moreover, within each group, ANOVA was performed to check the effect of storage time. When the differences were statistically significant ($p < 0.05$), a Tukey's test was carried out to identify differences between pairs of groups. Correlation between safranal and the sensorial and physicochemical parameters was determined by estimating Pearson correlation coefficients.

3. Results

3.1. Physicochemical Quality (pH and Color Parameters)

Table 1 shows the pH and color parameters of each group (control, A: 0.055%, B: 0.035%, and C: 0.015% w/w), and the changes in these values in the dry-cured ham slices during the storage period (0, 7, 14, 28, and 60 days). Throughout the storage period, pH values ranged from 5.96 to 5.42. From day 7, there were no significant differences among groups. In all samples, a gradual decrease in pH was observed with storage time, with significant differences between groups from 28 days of storage.

L* values were similar in all groups, and no statistical differences were found among samples at any time during storage. Notably, both in the control samples and in the flavored sample with the lowest saffron concentration (C), this parameter gradually decreased until 28 days of storage and then increased significantly. However, the L* values showed high stability in the A and B groups.

Redness, yellowness and Chroma did not vary with storage time. However, there was a significant difference due to the added saffron concentration at 28 days ($p < 0.01$). At this time point, the values of these color parameters followed the next order $A \geq B \geq C \geq$ control. In Hue (h°) these differences ($p < 0.01$) were observed at 28 days and at the end of the experiment and with the same above order. Huge angle showed a high stability in control and A groups. Visual appearance of the samples in each group during the storage time period is showed in Figure 1.

Table 1. Effect of different added concentrations of saffron and storage period on the physicochemical characteristics (pH and color; mean ± s.e.) of sachets of ham slices of ham.

Parameters	Storage Period (Days)	Concentration				ANOVA
		CONTROL (n = 5)	A (n = 5)	B (n = 5)	C (n = 5)	
pH	0	5.96 ± 0.12 ^{y,c}	5.95 ± 0.13 ^{xy,c}	5.77 ± 0.05 ^{x,b}	5.80 ± 0.10 ^{xy,b}	*
	7	5.76 ± 0.16 ^{b}	5.70 ± 0.03 ^{ab}	5.73 ± 0.03 ^{b}	5.68 ± 0.04 ^{b}	NS
	14	5.71 ± 0.04 ^{b}	5.73 ± 0.23 ^{bc}	5.71 ± 0.07 ^{b}	5.83 ± 0.25 ^{b}	NS
	28	5.69 ± 0.09 ^{b}	5.67 ± 0.02 ^{ab}	5.77 ± 0.07 ^{b}	5.69 ± 0.04 ^{b}	NS
	60	5.48 ± 0.05 ^{a}	5.48 ± 0.06 ^{a}	5.42 ± 0.05 ^{a}	5.44 ± 0.07 ^{a}	NS
Effect of storage period		***	***	***	***	
L*	0	42.37 ± 5.08 ^{ab}	45.05 ± 1.77	44.17 ± 7.08	47.28 ± 4.25 ^{b}	NS
	7	48.65 ± 5.22 ^{b}	49.15 ± 3.49	45.67 ± 3.44	45.02 ± 1.61 ^{b}	NS
	14	41.64 ± 4.61 ^{ab}	43.23 ± 5.67	45.54 ± 8.58	43.81 ± 4.61 ^{ab}	NS
	28	35.32 ± 1.95 ^{a}	40.94 ± 5.08	36.97 ± 4.63	35.55 ± 3.38 ^{a}	NS
	60	43.61 ± 3.73 ^{b}	45.45 ± 5.64	46.39 ± 6.34	46.09 ± 7.93 ^{b}	NS
Effect of storage period		**	NS	NS	**	
a*	0	19.41 ± 2.98	23.83 ± 3.20	22.15 ± 5.93	18.89 ± 3.50	NS
	7	19.22 ± 1.63	20.38 ± 1.82	21.67 ± 2.57	20.68 ± 1.69	NS
	14	18.32 ± 2.97	20.00 ± 4.39	18.96 ± 5.73	20.48 ± 4.11	NS
	28	17.85 ± 0.77 ^{x}	22.29 ± 2.21 ^{y}	22.28 ± 1.73 ^{y}	20.80 ± 2.38 ^{xy}	**
	60	18.97 ± 2.10	17.13 ± 5.53	15.45 ± 7.34	16.24 ± 7.26	NS
Effect of storage period		NS	NS	NS	NS	
b*	0	21.34 ± 7.85	29.27 ± 7.26	23.38 ± 5.81	18.70 ± 3.74	NS
	7	22.99 ± 6.48	31.62 ± 9.69	28.38 ± 3.84	24.91 ± 4.23	NS
	14	20.47 ± 7.69	30.63 ± 6.65	30.65 ± 14.02	25.73 ± 10.12	NS
	28	12.37 ± 1.59 ^{x}	28.40 ± 10.38 ^{y}	20.93 ± 4.27 ^{xy}	16.86 ± 6.32 ^{xy}	**
	60	16.23 ± 6.52	29.83 ± 4.73	28.91 ± 10.23	26.20 ± 9.66	NS
Effect of storage period		NS	NS	NS	NS	
Chroma (C*)	0	29.00 ± 7.67	37.84 ± 7.32	32.66 ± 5.64	26.71 ± 4.22	NS
	7	30.22 ± 5.07	37.80 ± 8.95	35.73 ± 4.42	32.45 ± 3.79	NS
	14	27.63 ± 7.53	37.00 ± 4.98	36.62 ± 13.29	33.13 ± 9.94	NS
	28	21.75 ± 0.99 ^{x}	36.37 ± 9.36 ^{y}	30.65 ± 3.88 ^{xy}	27.01 ± 5.46 ^{xy}	**
	60	25.41 ± 4.39	34.68 ± 5.31	32.91 ± 12.16	31.33 ± 10.32	NS
Effect of storage period		NS	NS	NS	NS	
Hue (h*)	0	46.40 ± 6.72	50.35 ± 4.77	46.46 ± 10.99 ^{a}	44.61 ± 6.02 ^{ab}	NS
	7	49.23 ± 8.10	56.19 ± 5.79	52.61 ± 2.12 ^{ab}	50.02 ± 4.39 ^{ab}	NS
	14	34.66 ± 3.87	50.28 ± 8.06	42.86 ± 4.44 ^{ab}	37.95 ± 7.81 ^{ab}	NS
	28	46.81 ± 7.01 ^{x}	56.26 ± 9.81 ^{y}	55.78 ± 12.52 ^{xy,a}	49.48 ± 8.86 ^{x,a}	**
	60	39.18 ± 11.75 ^{x}	60.49 ± 8.37 ^{y}	63.29 ± 6.48 ^{y,b}	57.60 ± 11.95 ^{y,b}	**
Effect of storage period		NS	NS	**	*	

CONTROL: sample without saffron; A: 0.055% *w*/*w*; B: 0.035% *w*/*w*; C: 0.015% *w*/*w*. NS: No significant. *, **, ***, indicates significance levels at 0.05, 0.01 and 0.001, respectively. ^{x,y}, values in the same row with different superscript are significantly different due to the group (CONTROL, A, B and C). ^{a,b,c}, values in the same column with different superscript are significantly different due to the different storage period (0, 7, 14, 28 and 60 days). s.e.: standard error.

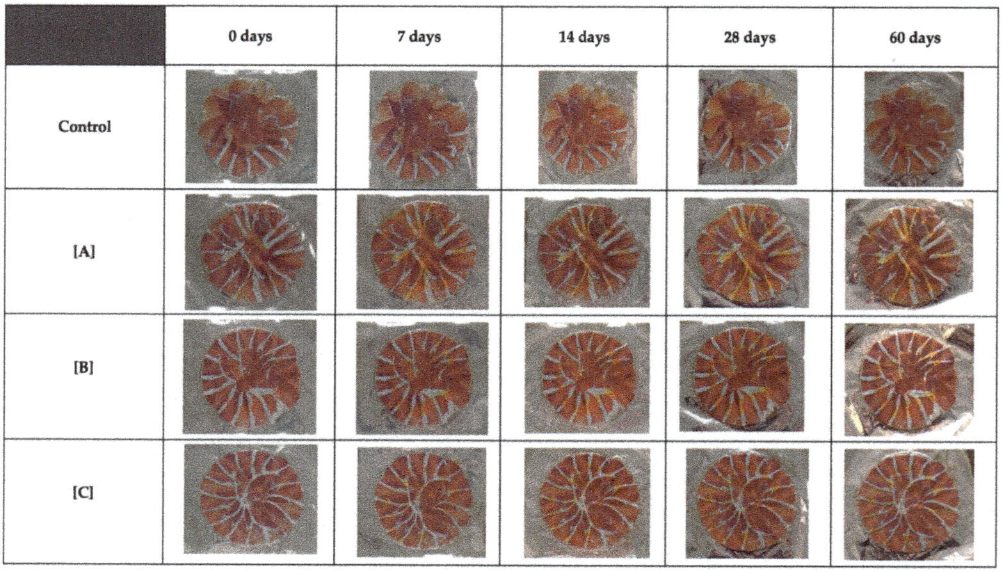

Figure 1. Visual appearance of the samples in each group (CONTROL: sample without saffron; A: 0.055% *w/w*; B: 0.035% *w/w*; C: 0.015% *w/w*) during the storage time period (0, 7, 14, 28 and 60 days). Scale (1:20).

3.2. Sensorial Quality

Table 2 shows the score given by the panelists to the spiced samples from day 7 of storage to the end of the experiment (60 days). There were no differences due to the gender of the panelists, and the results showed great stability in all groups, with values always higher than 3 and close to 4, which indicate that the spiced ham is to the taste (visual appearance, odor, and flavor) of consumers.

Table 2. Effect of different added concentrations of saffron on sensory characteristics (visual appearance, odor, and flavor; means ± s.e.) of sachets of ham slices.

Parameters	Storage Period (Days)	Concentration			ANOVA
		A (n = 30)	B (n = 30)	C (n = 30)	
Visual appearance	7	3.62 ± 0.85	3.74 ± 1.05	3.68 ± 0.98	NS
	14	3.42 ± 1.02 ^{x}	3.71 ± 0.92 ^{xy}	3.98 ± 0.84 ^{y}	*
	28	3.71 ± 1.02	3.66 ± 0.97	3.91 ± 0.92	NS
	60	3.75 ± 0.84	4.00 ± 0.76	3.53 ± 0.97	NS
Effect of storage period		NS	NS	NS	
Odor	7	3.76 ± 0.82	3.76 ± 1.01	3.68 ± 0.81	NS
	14	3.53 ± 0.95 ^{x}	3.80 ± 0.78 ^{xy}	3.94 ± 0.87 ^{y}	*
	28	3.60 ± 0.81 ^{x}	3.57 ± 1.09 ^{x}	4.17 ± 0.86 ^{y}	*
	60	3.42 ± 1.27 ^{x}	4.14 ± 0.87 ^{y}	4.08 ± 0.77 ^{y}	**

Table 2. *Cont.*

Parameters	Storage Period (Days)	Concentration			ANOVA
		A (n = 30)	B (n = 30)	C (n = 30)	
Effect of storage period		NS	NS	NS	
Flavor	7	3.50 ± 1.14	3.71 ± 0.87	3.79 ± 0.91	NS
	14	3.46 ± 1.08 ^{x}	4.00 ± 0.82 ^{y}	4.20 ± 0.72 ^{y}	***
	28	3.74 ± 0.95	3.91 ± 1.01	3.91 ± 0.78	NS
	60	3.58 ± 0.94	4.00 ± 0.86	3.72 ± 0.85	NS
Effect of storage period		NS	NS	NS	

A: 0.055% *w*/*w*; B: 0.035% *w*/*w*; C: 0.015% *w*/*w*. NS: No significant. *, **, *** indicates significance levels at 0.05, 0.01 and 0.001, respectively. ^{x,y}, values in the same row with different superscript are significantly different. 1: Do not like it; 2: I slightly dislike it; 3: Neither like nor dislike; 4: Like it; 5: I like it very much. s.e.: standard error.

Significant differences due to saffron concentration were observed in visual appearance ($p < 0.05$) and flavor ($p < 0.001$) at 14 days and in odor at 14, 28, and 60 days ($p < 0.05$ at 14 and 28 d; $p < 0.01$ at 60 days) with a similar trend for the three sensory parameters: C ≥ B ≥ A, depending on the time of analysis and the parameter. Figure 2 presents the percentage of panelists who considered a particular group favorite.

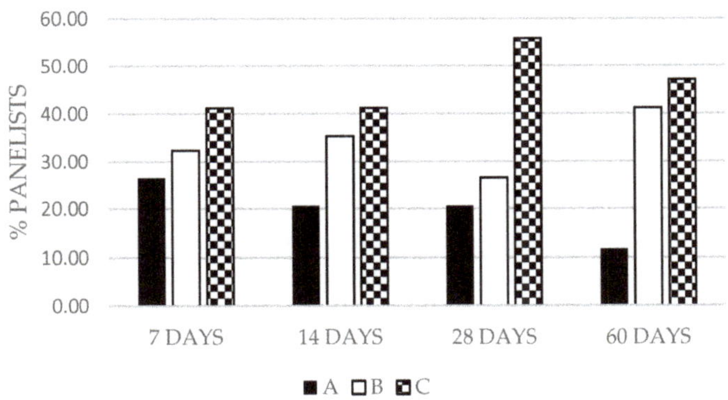

Figure 2. Percentage of panelists who considered a particular group favorite (A: 0.055% *w*/*w*; B: 0.035% *w*/*w*; C: 0.015% *w*/*w*) during the study period (7, 14, 28 and 60 days of storage).

3.3. *Transfer of Aromatic Compounds from Saffron*

Safranal content and its trend throughout the experiment is shown in Table 3. Only at 14 and 60 days of storage, there were differences ($p < 0.001$) among groups, and the groups were ordered according to safranal content as A > B > C and A > B = C, respectively. Safranal content decreased from day 7 in all samples, but subsequently, the tendency was different in each group with significant differences ($p < 0.001$ in A and B; $p < 0.05$ in C). It is noteworthy that safranal content increased in all samples at 60 days. Correlation between safranal content and other parameters is shown in Table 4. Only there was a significant correlation ($p < 0.05$) with odor ($r = 0.65$) in group C.

Table 3. Determination of safranal (µg/ 100 g ham; means ± s.e.) in sachets of sliced dry-cured ham during the storage period.

Parameter	Storage Period (Days)	Concentration			ANOVA
		A (n = 3)	B (n = 3)	C (n = 3)	
Safranal	7	4.45 ± 0.31 ^{b}	4.03 ± 0.18 ^{b}	3.22 ± 0.40 ^{b}	NS
	14	2.99 ± 0.12 $^{z, a}$	1.36 ± 0.01 $^{y, a}$	0.69 ± 0.07 $^{x, a}$	***
	28	2.89 ± 0.25 ^{a}	1.33 ± 0.29 ^{a}	3.00 ± 0.74 ^{b}	NS
	60	5.46 ± 0.27 $^{y, b}$	2.08 ± 0.46 $^{x, a}$	2.57 ± 0.26 $^{x, ab}$	***
Effect of storage period		***	***	*	

A: 0.055% *w/w*; B: 0.035% *w/w*; C: 0.015% *w/w* saffron. NS: No significant. *, ***, indicates significance levels at 0.05 and 0.001, respectively. ^{x,y,z}, values in the same row with different superscript are significantly different due to saffron concentration of saffron. ^{a,b}, values in the same column with different superscript are significantly different for the different storage period (7, 14, 28 and 60 days). s.e.: standard error.

Table 4. Correlation coefficients between safranal content with the sensorial and physicochemical parameters in each group.

Group	Visual Appearance	Flavor	Odor	pH	L*	a*	b*	C*	h*
A	−0.13	−0.25	−0.16	−0.57	0.27	−0.26	0.06	0	0.24
B	0.12	−0.11	0.45	−0.03	0.31	0.12	0.53	0.4	0.35
C	0.3	0.34	0.65 *	−0.49	−0.18	0.18	−0.11	−0.02	−0.15

A: 0.055% *w/w*; B: 0.035% *w/w*; C: 0.015% *w/w*. *, indicates correlation significant at the 0.05 level.

4. Discussion

4.1. Physicochemical Quality (pH and Color Parameters)

4.1.1. pH

The pH values found in our study were similar to those reported in previous studies [39,40] in dry ham after a similar storage time. The decline in this parameter is in agreement with the findings of a previous study [41] on the effect of storage under vacuum conditions for 8 months on dry-cured ham quality. In contrast, another study [39] on the shelf life of sliced dry-cured ham packaged under vacuum with analysis performed in the same storage period as the present study reported an increase in pH during storage time. This increase has been associated with the release of amino acids and other basic compounds during the dry-maturation stage [42].

pH is an important factor influencing the growth of microorganisms, with low pH inhibiting the growth of pathogens [43]. However, there are pathogenic microorganisms such as *Listeria monocytogenes* that can grow in the pH range observed in this study affecting the ham quality [44]. Therefore, other factors may be crucial to prevent their growth, such as low water activity and maintaining sliced ham at refrigeration temperatures [45]. According to [46] for cured meat product, such as ham, to be considered stable during storage and distribution, one of the conditions is that the pH is less than 6.0. In our study, this parameter was lower than this limit in all groups.

4.1.2. Color Parameters

Color is an important quality characteristic that contributes to the sensorial acceptability of dry-cured ham [47]. However, color is affected by many factors such as spices added, packaging or processing [48]. Changes in color parameters have been studied in dry ham [40,49–52].

The L* parameter has been associated with the thin layer of moisture on the muscle surface [53] and lightness in these muscles depends on the water content (moisture) and water movement (dehydration) towards the surface [42]. For some authors [47] is considered the most important parameter determining quality of meat products. According to [54] changes in this parameter in the sliced dry-cured ham could be negative since modifications in the typical color of dry-cured ham could influence consumers. It is evident that the addition of saffron with concentrations such as in A or B groups caused a high stability in lightness. Nevertheless, the results in C and control group were contrary to the results of authors such as [40,50] who determined that L* preserved color during similar storage time in sliced dry-cured ham.

According to [49], redness is used as an indicator of color stability while yellowness has been associated with rancidity. Authors such as [55] concluded that a* value was the most important aspect of color. In our study, the a* and b* parameters did not vary with storage time. Others [49] have also reported similar stability in a* and b* in ham slices after 8 weeks of storage in vacuum packaging. The obtained results could be attributed to the presence of crocetin esters, also known as crocins, a group of water-soluble carotenoids responsible for saffron's color strength [23,28,56]. Crocetin is formed from crocins during storage time [57]. Due to the fact crocetin is fat-soluble, it could cause the yellowness to increase in dry-cured ham slices [58].

According to [59] the characterization of the color is achieved by means of the coordinates of L*, a* and b*, but the main purpose in the measurements of the color is the objective determination of their differences through the parameters of chroma (C*) and of the tone (h*). Our results showed that the addition of saffron gives a greater C* and tone to the ham slices, reaching significant differences among groups at 28 days in both parameters and at the end of experiment in hue.Authors such as [60] have studied the chroma and hue in Spanish saffron and dry-cured Duroc ham [61], but there are not previous references which had studied the color parameters of dry-cured ham flavored with this spice.

4.2. Sensorial Quality

Sensory evaluation started developing with the growth of industry and processed food [62]. Sensory characteristics are crucial in the development of new food products [63] and influence consumer acceptance both before purchase (visual appearance) and at the time of consumption (odor and flavor). Because of this, sensory analysis are one of the most important methods in judging food quality [64]. Previous studies [52,65–70] have reported these parameters in ham and indicated the importance of flavor in the overall quality of dry-cured ham. However, the present study is the first to our knowledge to determine the degree of satisfaction of cured ham spiced with saffron. The addition of spices provide new tastes, colors and aromas to food that even gives culinary identity [71], owing to the changes in the composition of volatile compounds [72] that affect the hedonic characteristics [73] and may affect the acceptance of new products [74]. On the other hand, spices could improve the quality of meat products due to their preservatives properties [75].

The addition of saffron provoked a great stability during time of study in each group. In this work, all groups were accepted by consumers. It is evident that the panelists preferred the group with the lowest concentration of saffron (Figure 2). Other studies [52,76] that indicated the acceptability of dry-cured ham during storage obtained lower scores with storage time, owing to increasing rancid odor and flavor in vacuum-packed ham [50,51,77]. Despite the fact the shelf life assigned to Spanish dry-cured ham is approximately one year, this is significantly reduced when the dry-cured ham is sliced and vacuum-packaged [51,78]. The decrease of flavor, odor and even color is in accordance with the reduction of shelf life of ham, not due to microbiological problems, but because of the decrease of sensorial quality [51]. This rancidity is usually associated with a decrease in pH [79] and especially in products rich in unsaturated fatty acids [80] such as ham. This may have occurred in the present study (note that we did not analyze lipid oxidation) and affect the scores of the panelists. However, these scores did not vary significantly during the experimental period, a finding that may be attributed to the addition of saffron, which may have masked the negative effect of lipid oxidation or decelerate it owing to its antioxidant power [81]. Significant differences due to saffron concentration could be attributed that safranal, the major aromatic component of saffron, changes over time increasing its concentration [82,83] affecting to hedonic characteristics.

4.3. Transfer of Aromatic Compounds from Saffron

Because safranal is one of the major components of saffron [84] and represents 72% of the flavoring composition of saffron [85,86], its content was determined to assess the transfer of aromatic compounds from the spice to the ham. Such saffron compounds were not found in the control group ham samples, which indicates that dry-cured ham and saffron do not have common aromatics. The amount of safranal contained in dry-cured ham was much lower (10^{-7}) than the safranal content present in the spice itself [85]. This gives a subtle saffron flavor to the dry-cured ham without masking its origin flavor but enhancing it [28].

In all groups, there is a rapid decrease of 7 to 14 days, consistent with that detected by the panelists as shown in Table 2, and a different increase at each group to 60 days. These findings are consistent with previous findings [23,85] that indicated that safranal concentration is higher in saffron stored longer than a month because of formation of safranal from crocetin esters and picrocrocin during storage [28]. Previous studies reported that the main compounds of saffron change over time [82,83].

The method used to determine the transfer [22] only analyses the safranal in the surface layer of the slice. Therefore, as fat is a lipophilic medium that absorbs apolar substances [85], it causes a decrease in the safranal content of the such layers. However, during storage, the generation and the absorption of safranal compete, being the absorption process faster than the generation process. It could be due to the fact that the internal layers that have absorbed safranal became saturated with the compound generated after 28 days. This could occupy the surface layers, recovering the initial values of 7 days. It is shown in the evolution of A and C and the trend in B groups. Moreover, this is in agreement with the significant differences between groups at day 14 (with lower concentration of safranal) and 60 days (with higher concentration of safranal).

With decreasing concentration of saffron used to season the ham, the sensory scores improved, and the correlation changed from negative in group A to positive in group C ($r = 0.30$ with visual appearance, $r = 0.34$ with flavor, and $r = 0.65$ ($p < 0.05$) with odor). This agrees with the highest organoleptic scores obtained by group C samples (Table 2). Correlation between color parameters agrees with the previous results explained in Table 1, due to the change of the main saffron compounds during storage. Correlation of safranal content with pH (always negative) and with the color coordinates were not significant in any group.

5. Conclusions

The results of this study suggest that (1) the pH of ham decreases throughout storage, and (2) the color coordinates do not change over time even with the addition of saffron. (3) It is advisable not to use a saffron concentration higher than that used in group C because it negatively affects sensory acceptance. (4) The safranal content varies throughout storage and shows a positive correlation with sensory parameters, especially when saffron concentrations are lower. Future studies should analyze the effect of adding other spices to ham slices, to offer new meat products to consumers.

Author Contributions: Conceptualization, E.M.G.-S., N.M.-L., H.V. and G.L.A.; methodology, E.M.G.-S., N.M.-L., H.V. and G.L.A.; software, E.M.G.-S., N.M.-L.; validation, E.M.G.-S, N.M.-L., and G.L.A.; formal analysis, E.M.G.-S; investigation, E.M.G.-S., H.V., N.M.-L., and G.L.A.; resources, E.M.G.-S, H.V. and G.L.A.; data curation, E.M.G.-S., and N.M.-L.; writing—original draft preparation, E.M.G.-S., N.M.-L., H.V., and G.L.A.; writing—review and editing, E.M.G.-S., N.M.-L., H.V., and G.L.A.; visualization, E.M.G.-S., N.M.-L., H.V. and G.L.A.; supervision, H.V. and G.L.A.; project administration, H.V. and G.L.A.; funding acquisition, G.L.A. All authors have read and agreed to the published version of the manuscript.

Funding: CDTI FEDER (Ref: IDI-20180914).

Institutional Review Board Statement: Not applicable.

Informed Consent Statement: Not applicable.

Data Availability Statement: The data presented in this study are available from the corresponding author upon request.

Acknowledgments: Many thanks for the financial support given by Benibaldo, S.A.U. and CDTI-FEDER (Ref: IDI-20180914). The authors would like to thank to K. Walsh for assistance with the preparation of this manuscript in English.

Conflicts of Interest: The authors declare no conflict of interest.

References

1. FAOSTAT. Available online: http://www.fao.org/faostat/es/#data/QL (accessed on 30 January 2021).
2. Consumo, Comercialización y Distribución Agroalimentaria. Available online: https://www.mapa.gob.es/es/alimentacion/temas/consumo-tendencias/informe2019_vf_tcm30-540250.pdf (accessed on 28 July 2020).
3. Del Olmo, A.; Calzada, J.; Nuñez, M. Effect of High Pressure Processing and Modified Atmosphere Packaging on the Safety and Quality of Sliced Ready-to-Eat "Lacón", a Cured–Cooked Pork Meat Product. *Innov. Food Sci. Emerg. Technol.* **2014**, *23*, 25–32. [CrossRef]
4. Gallardo, E.; Narváez-Rivas, M.; Pablos, F.; Jurado, J.M.; León-Camacho, M. Subcutaneous Fat Triacylglycerols Profile from Iberian Pigs as a Tool To Differentiate between Intensive and Extensive Fattening Systems. *J. Agric. Food Chem.* **2012**, *60*, 1645–1651. [CrossRef]
5. Purslow, P.P. Chapter 1—Introduction. In *New Aspects of Meat Quality*; Woodhead Publishing Series in Food Science, Technology and Nutrition; Purslow, P.P., Ed.; Woodhead Publishing: Cambridge, UK, 2017; pp. 1–9, ISBN 978-0-08-100593-4.
6. Kim, J.; Yim, D.-G. Assessment of the Microbial Level for Livestock Products in Retail Meat Shops Implementing HACCP System. *Korean J. Food Sci. Anim. Resour.* **2016**, *36*, 594–600. [CrossRef]
7. Andrés, A.I.; Cava, R.; Ventanas, J.; Muriel, E.; Ruiz, J. Lipid Oxidative Changes throughout the Ripening of Dry-Cured Iberian Hams with Different Salt Contents and Processing Conditions. *Food Chem.* **2004**, *84*, 375–381. [CrossRef]
8. Tomažin, U.; Škrlep, M.; Prevolnik Povše, M.; Batorek Lukač, N.; Karolyi, D.; Červek, M.; Čandek-Potokar, M. The Effect of Salting Time and Sex on Chemical and Textural Properties of Dry Cured Ham. *Meat Sci.* **2020**, *161*, 107990. [CrossRef] [PubMed]
9. Fuentes, V.; Ventanas, J.; Morcuende, D.; Estévez, M.; Ventanas, S. Lipid and Protein Oxidation and Sensory Properties of Vacuum-Packaged Dry-Cured Ham Subjected to High Hydrostatic Pressure. *Meat Sci.* **2010**, *85*, 506–514. [CrossRef] [PubMed]
10. Coll-Brasas, E.; Gou, P.; Arnau, J.; Olmos, A.; Fulladosa, E. Processing Parameters Involved in the Development of Texture and Tyrosine Precipitates in Dry-Cured Ham: Modelisation of Texture Development. *Meat Sci.* **2021**, *172*, 108362. [CrossRef]
11. García-Díez, J.; Alheiro, J.; Pinto, A.L.; Soares, L.; Falco, V.; Fraqueza, M.J.; Patarata, L. Behaviour of Food-Borne Pathogens on Dry Cured Sausage Manufactured with Herbs and Spices Essential Oils and Their Sensorial Acceptability. *Food Control* **2016**, *59*, 262–270. [CrossRef]
12. El-Sayed, S.M.; Youssef, A.M. Potential Application of Herbs and Spices and Their Effects in Functional Dairy Products. *Heliyon* **2019**, *5*, e01989. [CrossRef]
13. Casquete, R.; Benito, M.J.; Martín, A.; Ruiz-Moyano, S.; Aranda, E.; Córdoba, M.G. Microbiological Quality of Salchichón and Chorizo, Traditional Iberian Dry-Fermented Sausages from Two Different Industries, Inoculated with Autochthonous Starter Cultures. *Food Control* **2012**, *24*, 191–198. [CrossRef]
14. Menéndez, R.A.; Rendueles, E.; Sanz, J.J.; Santos, J.A.; García-Fernández, M.C. Physicochemical and Microbiological Characteristics of Diverse Spanish Cured Meat Products. *CyTA J. Food* **2018**, *16*, 199–204. [CrossRef]
15. Mobasseri, M.; Ostadrahimi, A.; Tajaddini, A.; Asghari, S.; Barati, M.; Akbarzadeh, M.; Nikpayam, O.; Houshyar, J.; Roshanravan, N.; Alamdari, N.M. Effects of Saffron Supplementation on Glycemia and Inflammation in Patients with Type 2 Diabetes Mellitus: A Randomized Double-Blind, Placebo-Controlled Clinical Trial Study. *Diabetes Metab. Syndr.* **2020**, *14*, 527–534. [CrossRef] [PubMed]
16. Pourmasoumi, M.; Hadi, A.; Najafgholizadeh, A.; Kafeshani, M.; Sahebkar, A. Clinical Evidence on the Effects of Saffron (*Crocus Sativus* L.) on Cardiovascular Risk Factors: A Systematic Review Meta-Analysis. *Pharmacol. Res.* **2019**, *139*, 348–359. [CrossRef]
17. Milajerdi, A.; Djafarian, K.; Hosseini, B. The Toxicity of Saffron (*Crocus Sativus* L.) and Its Constituents against Normal and Cancer Cells. *J. Nutr. Intermed. Metab.* **2015**, *3*. [CrossRef]
18. Bagur, M.J.; Alonso Salinas, G.L.; Jiménez-Monreal, A.M.; Chaouqi, S.; Llorens, S.; Martínez-Tomé, M.; Alonso, G.L. Saffron: An Old Medicinal Plant and a Potential Novel Functional Food. *Molecules* **2017**, *23*, 30. [CrossRef]
19. Sena-Moreno, E.; Alvarez-Ortí, M.; Serrano-Díaz, J.; Pardo, J.E.; Carmona, M.; Alonso, G.L. Olive Oil Aromatization with Saffron by Liquid-Liquid Extraction. *J. Food Sci. Technol.* **2018**, *55*, 1093–1103. [CrossRef]
20. Amanpour, A.; Sonmezdag, A.S.; Kelebek, H.; Selli, S. GC-MS-Olfactometric Characterization of the Most Aroma-Active Components in a Representative Aromatic Extract from Iranian Saffron (*Crocus Sativus* L.). *Food Chem.* **2015**, *182*, 251–256. [CrossRef]

21. Valle García-Rodríguez, M.; Serrano-Díaz, J.; Tarantilis, P.A.; López-Córcoles, H.; Carmona, M.; Alonso, G.L. Determination of Saffron Quality by High-Performance Liquid Chromatography. *J. Agric. Food Chem.* **2014**, *62*, 8068–8074. [CrossRef] [PubMed]
22. Gómez-Sáez, E.M.; Moratalla-López, N.; Lorenzo, C.; Vergara, H.; Alonso, G.L. Determination of Saffron Volatiles by HS-SBSE-GC in Flavored Cured Ham. *Molecules* **2021**, *26*, 2073. [CrossRef]
23. García-Blázquez, A.; Moratalla-López, N.; Lorenzo, C.; Salinas, M.R.; Alonso, G.L. Effect of *Crocus Sativus* L. Stigmas Microwave Dehydration on Picrocrocin, Safranal and Crocetin Esters. *Foods* **2021**, *10*, 404. [CrossRef]
24. Himeno, H.; Sano, K. Synthesis of Crocin, Picrocrocin and Safranal by Saffron Stigma-like Structures Proliferated in Vitro. *Agric. Food Chem.* **1987**, *51*, 2395–2400. [CrossRef]
25. Gout, B.; Bourges, C.; Paineau-Dubreuil, S. Satiereal, a *Crocus Sativus* L. Extract, Reduces Snacking and Increases Satiety in a Randomized Placebo-Controlled Study of Mildly Overweight, Healthy Women. *Nutr. Res.* **2010**, *30*, 305–313. [CrossRef]
26. Forouzanfar, F.; Asadpour, E.; Hosseinzadeh, H.; Boroushaki, M.T.; Adab, A.; Dastpeiman, S.H.; Sadeghnia, H.R. Safranal Protects against Ischemia-Induced PC12 Cell Injury through Inhibiting Oxidative Stress and Apoptosis. *Naunyn. Schmiedebergs Arch. Pharmacol.* **2021**, *394*, 707–716. [CrossRef]
27. Pitsikas, N. *Crocus Sativus* L. Extracts and Its Constituents Crocins and Safranal; Potential Candidates for Schizophrenia Treatment? *Molecules* **2021**, *26*, 1237. [CrossRef]
28. Carmona Delgado, M.; Zalacain Aramburu, A. *The Chemical Composition of Saffron: Color, Taste and Aroma*, 1st ed.; Bomarzo: Albacete, Spain, 2006; ISBN 84-85977-91-6.
29. Barani, A.; Tajik, H. Simultaneous Determination of Saffron and Synthetic Dyes in Ready-to-Cook Iranian Barbecued Chicken by HPLC. *Int. J. Food Prop.* **2019**, *22*, 1608–1614. [CrossRef]
30. Shakeri, M.; Tayer, A.H.; Shakeri, H.; Jahromi, A.S.; Moradzadeh, M.; Hojjat-Farsangi, M. Toxicity of Saffron Extracts on Cancer and Normal Cells: A Review Article. *Asian Pac. J. Cancer Prev.* **2020**, *21*, 1867–1875. [CrossRef] [PubMed]
31. Aktypis, A.; Christodoulou, E.D.; Manolopoulou, E.; Georgala, A.; Daferera, D.; Polysiou, M. Fresh Ovine Cheese Supplemented with Saffron (*Crocus Sativus* L.): Impact on Microbiological, Physicochemical, Antioxidant, Color and Sensory Characteristics during Storage. *Small Rumin. Res.* **2018**, *167*, 32–38. [CrossRef]
32. Licón, C.C.; Carmona, M.; Berruga, M.I. Volatile Compounds in Pressed Ewes' Milk Cheese with Saffron Spice (*Crocus Sativus* L.). *Int. J. Dairy Technol.* **2015**, *68*, 399–408. [CrossRef]
33. Gaglio, R.; Gentile, C.; Bonanno, A.; Vintaloro, L.; Perrone, A.; Mazza, F.; Barbaccia, P.; Settanni, L.; Di Grigoli, A. Effect of Saffron Addition on the Microbiological, Physicochemical, Antioxidant and Sensory Characteristics of Yoghurt. *Int. J. Dairy Technol.* **2018**, 72. [CrossRef]
34. Villalba, T. *Código de Protección y Bienestar Animal*; Boletín Oficial del Estado: Madrid, Spain, 2017; ISBN 978-84-340-2435-9.
35. Sinkinson, C. Chapter 7—Triangle Test. In *Discrimination Testing in Sensory Science*; Rogers, L., Ed.; Woodhead Publishing: Cambridge, UK, 2017; pp. 153–170, ISBN 978-0-08-101009-9.
36. Anzaldúa-Morales, A. *La Evaluación Sensorial de los Alimentos en la Teoría y la Práctica*, 1st ed.; ACRIBIA: Zaragoza, Spain, 1994; ISBN 84-200-0767-6.
37. 14:00-17:00 ISO 3632-1:2011. Available online: https://www.iso.org/cms/render/live/en/sites/isoorg/contents/data/standard/04/45/44523.html (accessed on 21 March 2021).
38. Master Color Control. Available online: https://www.mcolorcontrol.com/archivos/L10-001_Understand_Color_es.pdf (accessed on 24 June 2021).
39. Piras, F.; Fois, F.; Casti, D.; Mazza, R.; Consolati, S.G.; Mazzette, R. Shelf Life of Sliced Dry-Cured Ham Packaged Under Vacuum. *J. Food Process. Preserv.* **2016**, *40*, 1223–1228. [CrossRef]
40. Parra, V.; Viguera, J.; Sánchez, J.; Peinado, J.; Espárrago, F.; Gutierrez, J.I.; Andrés, A.I. Effect of Exposure to Light on Physico-Chemical Quality Attributes of Sliced Dry-Cured Iberian Ham under Different Packaging Systems. *Meat Sci.* **2012**, *90*, 236–243. [CrossRef] [PubMed]
41. Cilla, I.; Martínez, L.; Beltrán, J.A.; Roncalés, P. Effect of Low-Temperature Preservation on the Quality of Vacuum-Packaged Dry-Cured Ham: Refrigerated Boneless Ham and Frozen Ham Cuts. *Meat Sci.* **2006**, *73*, 12–21. [CrossRef]
42. Pérez-Alvarez, J.A.; Sayas-Barberá, M.E.; Fernández-López, J.; Gago-Gago, M.A.; Pagán-Moreno, M.J.; Aranda-Catalá, V. Chemical and Color Characteristics of Spanish Dry-Cured Ham at the End of the Aging Process. *J. Muscle Foods* **1999**, *10*, 195–201. [CrossRef]
43. Stollewerk, K.; Jofré, A.; Comaposada, J.; Arnau, J.; Garriga, M. The Effect of NaCl-Free Processing and High Pressure on the Fate of Listeria Monocytogenes and Salmonella on Sliced Smoked Dry-Cured Ham. *Meat Sci.* **2012**, *90*, 472–477. [CrossRef]
44. Garriga, M.; Grèbol, N.; Aymerich, M.T.; Monfort, J.M.; Hugas, M. Microbial Inactivation after High-Pressure Processing at 600 MPa in Commercial Meat Products over Its Shelf Life. *Innov. Food Sci. Emerg. Technol.* **2004**, *5*, 451–457. [CrossRef]
45. Sadeghi-Mehr, A.; Lautenschlaeger, R.; Drusch, S. Behavior of Salmonella Spp. and Listeria Monocytogenes throughout the Manufacture and Shelf-Life of Dry-Cured Formed Ham. *Food Control* **2016**, *64*, 22–28. [CrossRef]
46. *Informe Del Comité Científico de La Agencia Española de Seguridad Alimentaria y Nutrición (AESAN) En Relación al Efecto de La Reducción de La Sal En La Seguridad Microbiológica de Los Productos Cárnicos Curados*; AESAN: Madrid, Spain, 2010.
47. García-Esteban, M.; Ansorena, D.; Gimeno, O.; Astiasarán, I. Optimization of Instrumental Colour Analysis in Dry-Cured Ham. *Meat Sci.* **2003**, *63*, 287–292. [CrossRef]

48. American Meat Science Association. *Meat Color Measurements Guidelines*; American Meat Science Association: Illinois, IL, USA, 2012.
49. García-Esteban, M.; Ansorena, D.; Astiasarán, I. Comparison of Modified Atmosphere Packaging and Vacuum Packaging for Long Period Storage of Dry-Cured Ham: Effects on Colour, Texture and Microbiological Quality. *Meat Sci.* **2004**, *67*, 57–63. [CrossRef]
50. Parra, V.; Viguera, J.; Sánchez, J.; Peinado, J.; Espárrago, F.; Gutierrez, J.I.; Andrés, A.I. Modified Atmosphere Packaging and Vacuum Packaging for Long Period Chilled Storage of Dry-Cured Iberian Ham. *Meat Sci.* **2010**, *84*, 760–768. [CrossRef] [PubMed]
51. Ramírez, R.; Contador, R.; Ortiz, A.; García-Torres, S.; López-Parra, M.M.; Tejerina, D. Effect of Breed Purity and Rearing Systems on the Stability of Sliced Iberian Dry-Cured Ham Stored in Modified Atmosphere and Vacuum Packaging. *Foods* **2021**, *10*, 730. [CrossRef]
52. Cilla, I.; Martínez, L.; Beltrán, J.A.; Roncalés, P. Dry-Cured Ham Quality and Acceptability as Affected by the Preservation System Used for Retail Sale. *Meat Sci.* **2006**, *73*, 581–589. [CrossRef] [PubMed]
53. Hunt, M.C. Meat Color Measurements. In Proceedings of the 33rd Annual Reciprocal Meat Conference, Chicago, IL, USA, 22–25 June 1980.
54. Morales, R.; Guerrero, L.; Aguiar, A.P.S.; Guàrdia, M.D.; Gou, P. Factors Affecting Dry-Cured Ham Consumer Acceptability. *Meat Sci.* **2013**, *95*, 652–657. [CrossRef] [PubMed]
55. Dvorak, P.; Musilova, H.; Svarcova, I. On-Line Measurements of Colour of Pork. *Fleischwirtschaft* **2001**, *81*, 89–91.
56. García-Rodríguez, M.V.; Moratalla-López, N.; López-Córcoles, H.; Alonso, G.L. Saffron Quality Obtained Under Different Forcing Conditions, Considering Various Vegetative Stages of Corms. *Sci. Hortic.* **2021**, *277*, 109811. [CrossRef]
57. Siracusa, L.; Gresta, F.; Ruberto, G. Saffron (Crocus sativus L.) Apocarotenoids: A Review of their Biomolecular Features and Biological Activity Perspectives. In *Carotenoids: Properties, Effects and Diseases*; Yamagughi, M., Ed.; Nova Science Publ. Inc.: New York, NY, USA, 2011; Volume 6, pp. 145–178.
58. Srivastava, R.; Ahmed, H.; Dixit, R.K.; Dharamveer; Saraf, S.A. *Crocus Sativus* L.: A Comprehensive Review. *Pharmacogn. Rev.* **2010**, *4*, 200–208. [CrossRef] [PubMed]
59. Domínguez Gómez, M.J. *Efectos de la Materia Prima y el Proceso Secado-Maduración Sobre la Calidad del Jamón Curado*; Universitat Politècnica de València: Valencia, Spain, 2020.
60. Anuar, N.; Taha, R.M.; Mahmad, N.; Mohajer, S.; Che Musa, S.A.N.; Zainal Abidin, Z.H. Correlation of Colour, Antioxidant Capacity and Phytochemical Diversity of Imported Saffron by Principal Components Analysis. *Pigment Resin. Technol.* **2017**, *46*, 107–114. [CrossRef]
61. Seong, P.N.; Park, K.M.; Kang, S.M.; Kang, G.H.; Cho, S.H.; Park, B.Y.; Van Ba, H. Effect of Particular Breed on the Chemical Composition, Texture, Color, and Sensorial Characteristics of Dry-Cured Ham. *Asian Australas. J. Anim. Sci.* **2014**, *27*, 1164–1173. [CrossRef]
62. Ruiz-Capillas, C.; Herrero, A.M.; Pintado, T.; Delgado-Pando, G. Sensory Analysis and Consumer Research in New Meat Products Development. *Foods* **2021**, *10*, 429. [CrossRef]
63. Worch, T.; Lê, S.; Punter, P. How Reliable Are the Consumers? Comparison of Sensory Profiles from Consumers and Experts. *Food Qual. Prefer.* **2010**, *21*, 309–318. [CrossRef]
64. Djekic, I.; Lorenzo, J.M.; Munekata, P.E.S.; Gagaoua, M.; Tomasevic, I. Review on Characteristics of Trained Sensory Panels in Food Science. *J. Texture Stud.* **2021**. [CrossRef]
65. Ortiz, A.; Tejerina, D.; Díaz-Caro, C.; Elghannam, A.; García-Torres, S.; Mesías, F.J.; Trujillo, J.; Crespo-Cebada, E. Is Packaging Affecting Consumers' Preferences for Meat Products? A Study of Modified Atmosphere Packaging and Vacuum Packaging in Iberian Dry-Cured Ham. *J. Sens. Stud.* **2020**, *35*. [CrossRef]
66. Sugimoto, M.; Sugawara, T.; Obiya, S.; Enomoto, A.; Kaneko, M.; Ota, S.; Soga, T.; Tomita, M. Sensory Properties and Metabolomic Profiles of Dry-Cured Ham during the Ripening Process. *Food Res. Int.* **2020**, *129*. [CrossRef] [PubMed]
67. Díaz-Caro, C.; García-Torres, S.; Elghannam, A.; Tejerina, D.; Mesias, F.J.; Ortiz, A. Is Production System a Relevant Attribute in Consumers' Food Preferences? The Case of Iberian Dry-Cured Ham in Spain. *Meat Sci.* **2019**, *158*. [CrossRef] [PubMed]
68. Fernández, M.; Hospital, X.F.; Cabellos, C.; Hierro, E. Effect of Pulsed Light Treatment on Listeria Inactivation, Sensory Quality and Oxidation in Two Varieties of Spanish Dry-Cured Ham. *Food Chem.* **2020**, *316*. [CrossRef] [PubMed]
69. Schivazappa, C.; Virgili, R. Impact of Salt Levels on the Sensory Profile and Consumer Acceptance of Italian Dry-Cured Ham. *J. Sci. Food Agric.* **2020**, *100*, 3370–3377. [CrossRef] [PubMed]
70. Sirtori, F.; Dimauro, C.; Bozzi, R.; Aquilani, C.; Franci, O.; Calamai, L.; Pezzati, A.; Pugliese, C. Evolution of Volatile Compounds and Physical, Chemical and Sensory Characteristics of Toscano PDO Ham from Fresh to Dry-Cured Product. *Eur. Food Res. Technol.* **2020**, *246*, 409–424. [CrossRef]
71. De La Torre, J.E.; Gassara, F.; Kouassi, A.; Brar, S.; Belkacemi, K. Spice Use in Food: Properties and Benefits. *Crit. Rev. Food Sci. Nutr.* **2015**, *57*. [CrossRef]
72. Jung, S.; Jo, C.; Kim, I.S.; Nam, K.C.; Ahn, D.U.; Lee, K.H. The Influence of Spices on the Volatile Compounds of Cooked Beef Patty. *Korean J. Food Sci. Anim. Resour.* **2014**, *34*, 166–171. [CrossRef]

73. Dini, I. Chapter 14—Spices and Herbs as Therapeutic Foods. In *Food Quality: Balancing Health and Disease*; Handbook of Food Bioengineering; Holban, A.M., Grumezescu, A.M., Eds.; Academic Press: Cambridge, MA, USA, 2018; pp. 433–469, ISBN 978-0-12-811442-1.
74. Shahidi, F. Flavor of meat and meat products—An overview. In *Flavor of Meat and Meat Products*; Shahidi, F., Ed.; Springer: Boston, MA, USA, 1994; pp. 1–3, ISBN 978-1-4615-2177-8.
75. Gottardi, D.; Bukvicki, D.; Prasad, S.; Tyagi, A.K. Beneficial Effects of Spices in Food Preservation and Safety. *Front. Microbiol.* **2016**, *7*, 1394. [CrossRef]
76. Ruiz, J.; García, C.; Muriel, E.; Andrés, A.I.; Ventanas, J. Influence of Sensory Characteristics on the Acceptability of Dry-Cured Ham. *Meat Sci.* **2002**, *61*, 347–354. [CrossRef]
77. Armero, E.; Flores, M.; Toldrá, F.; Barbosa, J.-A.; Olivet, J.; Pla, M.; Baselga, M. Effects of Pig Sire Type and Sex on Carcass Traits, Meat Quality and Sensory Quality of Dry-Cured Ham. *J. Sci. Food Agric.* **1999**, *79*, 1147–1154. [CrossRef]
78. Kemp, J.D.; Langlois, B.E.; Akers, K.; Means, W.J.; Aaron, D.K. Effect of Storage Temperature and Time on the Quality of Vacuum Packaged Dry-Cured Ham Slices. *J. Food Sci.* **1988**, *53*, 402–406. [CrossRef]
79. Juncher, D.; Rønn, B.; Mortensen, E.; Henckel, P.; Karlsson, A.; Skibsted, L.; Bertelsen, G. Effect of Pre-Slaughter Physiological Conditions on the Oxidative Stability of Colour and Lipid during Chill Storage of Pork. *Meat Sci.* **2001**, *58*, 347–357. [CrossRef]
80. RESEARCHGATE. Available online: https://www.researchgate.net/publication/284179196_Mejora_de_las_caracteristicas_del_jamon_curado_a_traves_de_la_alimentacion (accessed on 28 June 2021).
81. Bagur, M.J.; Alonso Salinas, G.L.; Jiménez-Monreal, A.M.; Serrano-Heras, G.; Martínez-Tomé, M.; Alonso, G.L. Effect of Daily Intake of a Saffron Infusion on Blood Cholesterol Levels. *Appl. Sci.* **2020**, *10*, 4763. [CrossRef]
82. Carmona, M.; Zalacain, A.; Pardo, J.E.; López, E.; Alvarruiz, A.; Alonso, G.L. Influence of Different Drying and Aging Conditions on Saffron Constituents. *J. Agric. Food Chem.* **2005**, *53*, 3974–3979. [CrossRef]
83. Chaouqi, S.; Moratalla-López, N.; Lage, M.; Lorenzo, C.; Alonso, G.L.; Guedira, T. Effect of Drying and Storage Process on Moroccan Saffron Quality. *Food Biosci.* **2018**, *22*, 146–153. [CrossRef]
84. Alonso, G.L.; Salinas, M.R.; Sánchez-Fernández, M.A.; Garijo, J. Safranal Content in Spanish Saffron. *Food Sci. Technol. Int.* **2001**, *7*, 225–229. [CrossRef]
85. Carmona, M.; Zalacain, A.; Salinas, M.R.; Alonso, G.L. A New Approach to Saffron Aroma. *Crit. Rev. Food Sci. Nutr.* **2007**, *47*, 145–159. [CrossRef]
86. Alonso, G.L.; Salinas, M.R.; Esteban-Infantes, F.J.; Sánchez-Fernández, M.A. Determination of Safranal from Saffron (*Crocus Sativus* L.) by Thermal Desorption−Gas Chromatography. *J. Agric. Food Chem.* **1996**, *44*, 185–188. [CrossRef]

foods

MDPI

Article

Consumer Attitudes towards Technological Innovation in a Traditional Food Product: The Case of Wine

Adrián Rabadán

Higher Technical School of Agriculture and Forestry Engineering, University of Castilla-La Mancha, 02071 Albacete, Spain; Adrian.Rabadan@uclm.es

Abstract: Food innovation is crucial for food companies in order to produce healthier, safer, and more convenient foods. However, there is a segment of consumers reluctant to accept new foods. This attitude is even more important when those novelties are developed in products such as wine that have habitually relied on heritage and traditional production as their main competitive advantage. In this study, consumer attitudes toward innovation in the wine industry were evaluated by simultaneously considering product neophobia and process neophobia. Based upon a sample of 400 personal interviews with Spanish wine consumers, the results showed that these two aspects of neophobia were uncorrelated, meaning they are useful to measure different aspects of general food neophobia. Cluster analysis showed that four different segments of consumers exist, with different attitudes toward technological innovation in the wine industry. The consumer segment that shows the most positive attitudes toward wine innovation (product and process innovation) is that with the highest income and highest level of education. Moreover, greater involvement with the product (wine) results in lower product neophobia. Therefore, future studies should consider product involvement and exposure to cultural diversity as essential factors when evaluating food neophobia.

Keywords: product innovation; process innovation; neophobia; food technology neophobia; wine neophobia scale

Citation: Rabadán, A. Consumer Attitudes towards Technological Innovation in a Traditional Food Product: The Case of Wine. *Foods* **2021**, *10*, 1363. https://doi.org/10.3390/foods10061363

Academic Editor: Jean-Xavier Guinard

Received: 17 May 2021
Accepted: 9 June 2021
Published: 12 June 2021

1. Introduction

The increasing interest of food companies in innovation in the search for competitiveness has encouraged the development of innovations even in food sectors in which tradition has long been established as their main competitive advantage [1]. Few food sectors are as influenced by heritage and tradition as the wine production sector, especially in the traditional wine-growing countries [2]. However, changing consumer tastes and the emergence of nontraditional wine markets worldwide have encouraged wine makers to adapt their products in order to succeed [3–6]. However, innovation in the production of traditional food products must be evaluated carefully [1,7]. In a study conducted in different European countries, Kühne, Vanhonacker, Gellynck, and Verbeke [1] concluded that consumers were open to accepting innovation in traditional food products as long as the traditional character of those foods was preserved. As a result, even if significant consumers segments are interested in new wines [2,5], these innovations should be considered carefully.

Although innovation in the food sector is mainly devoted to the development of novel products [8], innovation within food companies should be understood as a wider process [9]. Drawing on the Oslo Manual [10], two different types of innovation can be identified: technological and nontechnological innovations. The former refers to the development of new products or production processes, while the latter refers to the innovations that companies implement in their marketing strategies or their management [11]. In the wine sector, technological innovations include a range of possibilities, from the addition of different extracts [5] to the use of fungus-resistant grapes [12] or changes in the winemaking process [13]. On the other hand, nontechnological innovations include changes in marketing strategies [14] or the implementation of voluntary certification schemes [15].

Although food innovation is crucial to meet the increasing demand of safer, healthier, and more convenient foods [16,17], the development of novel products faces the rejection of consumers that perceive risks or disadvantages in these innovative foods [18,19]. Several studies have reported that only a small percentage of launched novel food products succeed, while the vast majority continue to fail [7,20–22]. This negative perception increases when production is perceived to be more industrial [23] as food innovation is sometimes associated with processed products and unhealthiness [24]. With this in mind, the development of new food products must consider the effects of so-called Food Neophobia (FN), which was defined by Henriques et al. [25] as the personal reluctance to accept and/or enjoy new or unfamiliar foods.

Due to the direct effect of FN on the success of food innovation, numerous tools have been proposed to evaluate this attitude [26]. However, thirty years after its presentation, the most successful tool to evaluate FN is still the Food Neophobia Scale (FNS) developed by Pliner and Hobden [27]. This scale has been used as the basis for the development of other scales to measure additional aspects of FN. This is the case of the so-called Food Technology Neophobia Scale (FTNS), developed to evaluate consumers' reluctance toward foods produced using new technologies [28]. Various studies have reported that although the FNS and the FTNS are related, the low correlations found on the scores of the two scales (ranges from −0.12 to 0.33) [19] show that they measure different aspects of FN, supporting the idea of general FN as a complex variable that can only be evaluated using a combination of different instruments.

Wine is a complex product, and, for this reason, consumer wine consumer preferences are influenced by a wide range of factors from the emotional and social to intrinsic quality factors [29–31]. The importance attributed to each quality factor, from origin to wine barrel aging, depends on consumer involvement with the product [32] but also on individual attitudes that appear as a result of social background [5]. All these specific characteristics of wine encouraged the creation of a specific Wine Neophobia Scale (WNS) [6] using the existing FNS, to specifically evaluate wine neophobia (WN). Few studies have been developed using the WNS, although there are some noteworthy exceptions. This is the case of the study by Nguyen, Johnson, Jeffery, Danner, and Bastian [5], who, after evaluating the level of WN in Australia, Vietnam, and China, found that WN is country-dependent.

After thirty years of studies on FN, some general conclusions have been reached, while other areas are still open to debate [33]. Knaapila et al. [34] reported that up to two-thirds of the variation in individual FN was genetically determined. Supporting the idea of FN as a personal trait, lower FN scores have been associated with a higher individual openness [35] and greater exposure to cultural diversity [36]. Following this association, higher FN scores have also been reported for individuals living in rural areas [36,37]. In this regard, most of the studies coincide in reporting higher levels of FN in individuals with lower levels of education and income [37–42]. However, it is still unknown whether the general conclusions reached for general FN might also apply to WN.

The current study aimed to evaluate consumer attitudes toward innovation in the wine industry, including the combined study of WN and food technology neophobia (FTN). The identification and characterization of consumers with the most positive attitudes toward innovation is crucial to develop strategies that ensure wine companies committed to wine innovation achieve higher profits than penalties as a result of their breaking with the association between wine and tradition and heritage.

2. Materials and Methods

2.1. Database

The data used in this study were collected using survey personal interviews to consumers when they were about to shop in different establishments. The survey was administered in September 2019 in the cities of Madrid, Leganés, and Toledo (Spain) as representative examples of large, medium, and small-sized towns, respectively. A total of 400 individuals who reported to have drunk wine at least once in the previous month were interviewed. A filter question was used for this purpose. The socioeconomic characteristics of the consumers interviewed are shown in Table 1. The maximum sampling error was below 5.0% for a confidence level of 95.5% (k = 2) under the principle of maximum indetermination (p = q = 50%). To confirm that the survey questions were well designed and were easily understandable, the preliminary questionnaire was administered to 30 wine consumers before the fieldwork.

Table 1. Socioeconomic characteristics of the sample (%).

Variables	Percentage (%)
Gender	
Men	56.9
Women	43.1
Age (in years)	
18–24	14.0
25–34	21.0
35–49	20.0
50–64	25.0
>65	20.0
Education level	
Elementary	28.5
Secondary	39.5
University	30.6
Postgraduate	1.5
Net monthly family income	
<EUR 1500	17.5
EUR 1500–2100	22.8
EUR 2101–3000	35.6
>EUR 3000	24.2

2.2. *Methodology*

In order to evaluate consumer WN, the Wine Neophobia Scale (WNS) was administered [6] (Table 2). The WNS is an 8-item scale developed to evaluate consumer neophobia toward the acquisition of new wines, based on the well-known FNS originally developed by Pliner and Hobden [27] to evaluate general food neophobia. As the questionnaire was administered in Spanish, the translation proposed by Fernández-Ruiz et al. [43] was used, changing the word "*alimento*" (food) to "*vino*" (wine). In the WNS, consumers are asked to evaluate each of the proposed items on a 9-point Likert-type scale ranging from "strongly disagree" (1) to "strongly agree" (9). The individual score on the WNS for each consumer was calculated as the sum of the 8 items after reversing the negative items (items 2, 4, 5, and 8). As a result, consumers obtaining higher values indicate a higher level of wine neophobia and, as a result, a lower tendency to try new wines. On the other hand, consumers with lower scores are identified as wine neophilics, indicating a higher tendency to try new wines.

To evaluate consumer attitudes toward foods produced using novel technologies, the Abbreviated Food Technology Neophobia Scale (AFTNS) was used [44] (Table 2). The AFTNS is an abbreviated 9-item version of the original 13-item Food Technology Scale proposed by Cox and Evans [28]. The translation to Spanish provided by the study of Schnettler et al. [45] was used. Specifically, consumers were asked to evaluate the 9 proposed items using a 6-point Likert-type scale. Individual scores for each consumer were calculated as the sum of the scores for each item after reversing the negative item (item 5). Within the possible range of the scale (9 to 54), higher values indicate a higher level of food technology neophobia (FTN), meaning consumers with a low tendency to accept products that have been produced using novel technologies (NT).

Table 2. Items from the Wine Neophobia Scale (WNS) and the Abbreviated Food Technology Neophobia Scale (AFTNS).

WNS		AFTNS	
Item	**Statements**	**Item**	**Statements**
1	I like going to places serving wines from different countries. (R)	1	New foods are not healthier than traditional foods.
2	I will drink almost any wine. (R)	2	The benefits of new food technologies are often grossly overstated.
3	I am afraid to drink wines I have never had before.	3	There are plenty of tasty foods around, so we do not need to use new food technologies to produce more.
4	At social gatherings, I will try a new wine. (R)	4	New food technologies decrease the natural quality of food.
5	I like wines from different countries. (R)	5	New food technologies are unlikely to have long-term negative health effects. (R)
6	If I do not know what wine it is, I won't try it.	6	New food technologies may have long-term negative environmental effects.
7	I do not trust new wines.	7	It can be risky to switch to new food technologies too quickly.
8	I am constantly trying new and different wines. (R)	8	Society should not depend heavily on technologies to solve its food problems.
		9	There is no sense trying out high-tech food products because the ones I eat are good enough.

Additionally, consumers provided information about the importance they attached to different wine attributes (such as price or origin) using a 5-point Likert-type scale ranging from very unimportant (1), neither unimportant nor important (3), to very important (5). To identify the attributes included in the survey, previous studies evaluating factors that determine wine purchase decision-making were used [5,46]. Consumers were also required to indicate the frequency of their consumption of wine (four levels). Using a 9-point Likert-type scale from strongly disagree (1) to strongly agree (9), consumers were also asked to indicate their attitudes toward statements linked to wine innovation and their opinions on statements about individual openness. Moreover, to include information on the consumers' socioeconomic characteristics, they were asked to provide information about gender, age, highest level of education completed, and net household income.

Similarly to Schnettler, Grunert, Miranda-Zapata, Orellana, Sepúlveda, Lobos, Hueche, and Höger [44], a cluster analysis using hierarchical conglomerates was used with linkage by Ward's method and the squared Euclidian distance as the measure of similarity between objects [47]. The aim of the segmentation was to identify consumer segments according to consumer scores obtained on the WNS and AFTNS. Four different groups with different attitudes were obtained. To describe the differences between the obtained segments, one-way analysis of variance (ANOVA) with Tukey's HSD post-hoc (significance level 5%) comparison was used to examine responses about wine attributes and the results of the scales, and Pearson's chi^2 test was applied to discrete variables. For the statistical analysis of the data, the Statistical Package for Social Sciences IBM SPSS version 23 was used.

3. Results and Discussion

The results show that consumers interviewed in this study are more neophobic than the average consumer in Australia [5]. In this regard, the deeper tradition of Spain as a growing wine country could be the explanation for consumers' less positive attitude toward the acceptance of novel wines [2]. However, this explanation contradicts the widely accepted idea of FN decreasing with the exposure to product diversity [36]. In European countries in which viticulture is traditional, wine consumers are exposed from birth to a wide range of different wines, yet their level of wine neophobia seems to be higher than in other emerging markets such as Australia. Castellini and Samoggia [2] suggested that in European countries in which viticulture is traditional, consumers are influenced by their countries' wine traditional heritage but, at the same time, excited about trying novel or innovative wines. To our knowledge, only the study by Nguyen, Johnson, Jeffery,

Danner, and Bastian [5] has evaluated the level of wine neophobia in several countries simultaneously, concluding that consumers from Vietnam were more neophobic than those from China and Australia.

The correlation found between the score obtained by consumers on the WNS and the AFTNS was not significant, showing a value within the range reported for previous studies between the FNS and the FTNS [19,28,44,48–51]. This result confirms the proposed idea of the WNS and FTNS measuring different aspects of FN. Moreover, the average score obtained on the WNS for Spanish consumers is similar to the values reported by Fernández-Ruiz, Claret, and Chaya [43] for general FN using the FNS. In this sense, further studies of the correlation between WN and general FN should be conducted to evaluate whether one could be used as a proxy of the other.

The results of the consumer segmentation analysis using the scores obtained on the WNS and the AFTNS are shown in Table 3. Four different consumer segments comprising a similar number of consumers were obtained. Segment 1 is composed of those reported as wine neophilics as they show the lowest values of WN and TFN. A segment of neophobics was also identified (Segment 4). This segment was formed by consumers that showed the highest values of WN and FTN, meaning they had the lowest interest in trying new wines and also the lowest rate of acceptance of any product produced using innovative technologies. Two consumer segments with intermediate characteristics between these two groups were also identified. Segment 2 was formed by consumers with low values of WN but high values for TFN. Opposite attitudes were reported by those consumers aggregated in Segment 3.

Table 3. Consumer segmentation according to Wine Neophobia Scale (WNS) and Abbreviated Food Technology Scale (AFTNS).

	Segment 1 Neophilics (n = 102)	**Segment 2 Neophilics Anti-NT (n = 93)**	**Segment 3 Neophobics Pro-NT (n = 112)**	**Segment 4 Neophobics (n = 93)**
		Scales		
WNS (p = 0.000)	23.59 ^{b}	24.06 ^{b}	40.34 ^{a}	41.33 ^{a}
AFTNS (p = 0.000)	18.79 ^{c}	33.25 ^{a}	21.45 ^{b}	34.49 ^{a}
		Consumption frequency (%)		
A few times per week	20.6	21.5	9.8	18.3
Once per week	43.1	40.9	18.8	20.4
Once per two weeks	8.8	15.1	15.2	15.1
Once per month	27.5	22.6	26.3	46.2
		Wine attributes		
Type (p = 0.007)	4.32 ^{ab}	4.59 ^{a}	4.15 ^{b}	4.48 ^{a}
Grape variety (p = 0.000)	3.53 ^{a}	3.66 ^{a}	2.71 ^{b}	3.01 ^{b}
Barrel aging (p = 0.000)	4.09 ^{a}	4.01 ^{a}	2.95 ^{b}	3.31 ^{b}
Price (p = 0.068)	3.77	4.22	3.95	4.12
Origin (p = 0.001)	3.63 ^{ab}	3.88 ^{a}	3.31 ^{c}	3.32 ^{bc}
Brand image (p = 0.000)	3.94 ^{a}	3.90 ^{a}	3.27 ^{b}	3.56 ^{ab}
Bottle aesthetics (p = 0.010)	3.16 ^{ab}	3.30 ^{a}	2.71 ^{b}	2.85 ^{ab}
Quality labels (p = 0.001)	3.69 ^{a}	3.67 ^{a}	3.04 ^{b}	3.25 ^{b}
Organic (p = 0.000)	2.36 ^{a}	2.45 ^{a}	1.94 ^{b}	1.75 ^{b}

The means of the values for the consumers in each segment are shown for the scales and for the wine attributes. Different letters in the same row mean significant differences for scales and wine attributes ($p < 0.05$). Chi-squared values for wine consumption frequency are: $X^2 = 44.101$, df = 9, $p = 0.000$.

Segments composed of consumers that show lower WN scores (Segments 1 and 2) are also those that report a higher frequency of wine consumption (more than 60% of them consume wine at least once a week). This result supports the idea that a greater exposure to the product, in this case, a higher frequency of wine consumption, is associated with a lower product neophobia [36]. Similar results were obtained by Nguyen, Johnson, Jeffery, Danner, and Bastian [5], who also found a correlation between consumption frequency and WN in different countries. More involved consumers (the group of neophilics and neophilics anti-NT) attach greater importance to most of the wine attributes considered, including grape variety, wine barrel aging, or the presence of quality labels. When purchasing food products, a higher involvement has also been associated with the evaluation of a wider range of attributes in products such as fruits [52]. On the other hand, WN has no significant effect on the importance consumers attach to the price when purchasing wine.

Table 4 shows the attitudes of the reported segments toward the statements linked to the wine purchasing process. Wine neophilic segments showed more positive attitudes toward online wine shopping. Regarding wine distribution, Casali et al. [53] found that innovative wine companies opt more for direct distribution channels. Our results suggest that innovative consumers could shop at innovative wineries using online sales as a direct selling channel. Consumers that are more open to wine innovation also place more trust in wine advertising more and pay greater attention to bottle aesthetics and winery image when purchasing wine. This encourages wine makers to develop nontechnological innovations (marketing and organizational innovations) also directed to these segments of wine neophilics.

Table 4. Impacts of WN and FTN on wine consumers' opinions toward statements linked to wine innovation.

	Neophilics (n = 102)	**Neophilics Anti-NT (n = 93)**	**Neophobics Pro-NT (n = 112)**	**Neophobics (n = 92)**
I would like to buy more wine online (p = 0.000)	3.37 [a]	2.52 [b]	1.72 [c]	1.42 [c]
I would like to have my favorite wine accessible in more retailers (p = 0.088)	1.63	2.13	1.63	1.72
I trust advertising about new wines (p = 0.000)	7.08 [a]	6.75 [a]	5.13 [b]	5.54 [b]
Wine bottle aesthetics are very important to me (p = 0.000)	6.02 [a]	6.02 [a]	3.96 [c]	5.04 [b]
I value the corporate image of the winery when purchasing a wine (p = 0.000)	7.03 [a]	6.71 [a]	4.59 [c]	5.52 [b]

The means of the values for the consumers in each segment. Likert 9-point scale from strongly agree (1) to strongly disagree (9). Different letters in the same row mean significant differences in the consumers' opinions ($p < 0.05$).

Table 5 shows the socioeconomic characteristics of each designated segment. Significant differences were found between consumer segments regarding income and education, but not according to gender or age. No effect of gender on consumer neophobia was expected as, although a study developed in very specific populations reported some effects [54], large-scale studies developed in countries such as Portugal [55] or Sweden [56] have concluded that there is no association between gender and FN.

Although significant differences for age were not found, the age profile of the segment of neophilic consumers showed slight differences compared to the profile of neophobics. For example, the number of consumers over 65 years old was twice as large in the neophobic segment as in the neophilic segment. In this sense, Castellini and Samoggia [2] reported that in countries with a long tradition of wine consumption, young consumers show more positive attitudes toward wine innovation [2]. However, in one of the most comprehensive studies about evolution of FN with age, Dovey et al. [57] concluded that FN reaches its highest point in childhood and then decreases during adolescence, is stable during adulthood, and then increases slightly in older ages as health problems appear. As wine consumption is mainly concentrated in adulthood, the reported absence of age-related differences between segments is consistent with the model proposed by Dovey, Staples, Gibson, and Halford [57].

On the other hand, significant differences were found for the variables of education and income. The segment of neophilic consumers was composed of the consumers with the highest income and level of education, while the segment of neophobic consumers showed the lowest income and the lowest education level. Currently, there is a consensus on the negative association between individuals' FN scores and higher education level and income [33,37,40]. According to our results, this association between lower income and education and higher neophobia also appears when WN is considered instead of general FN. Flight, Leppard, and Cox [36] explained this association by considering that exposure to greater cultural diversity is expected for consumers with a higher socio-economic status, which results in lower neophobia.

Table 5. Socioeconomic characteristics of the segments (%).

	Neophilics (n = 102)	Neophilics Anti-NT (n = 92)	Neophobics Pro-NT (n = 112)	Neophobics (n = 92)
		Gender (%)		
Men	65.7	57	50	55.9
Women	34.3	43	50	44.1
		Age (%)		
18–24	11.8	11.8	17	15.1
25–34	20.6	23.7	20.5	19.4
35–46	25.5	22.6	17	15.1
50–64	28.4	23.7	24.1	23.7
65+	13.7	18.3	21.4	26.9
		Education level (%)		
Elementary	20.6	29	28.6	36.6
Secondary	41.2	35.5	48.2	31.2
University	34.3	33.3	23.2	32.3
Postgraduate	3.9	2.2	0	0
		Net family income (%)		
<EUR 1500	9.3	33.8	7.2	21.9
EUR 1500–2100	13.5	9.3	38.1	29.3
EUR 2101–3000	39.6	32.6	41.2	28
>EUR 3000	37.5	24.4	13.4	20.7

Chi-squared values for the socioeconomic variables are: gender, $X^2 = 5.42$, df = 3, $p = 0.143$; age, $X^2 = 26.79$, df = 12, $p = 0.571$; education level, $X^2 = 18.703$, df = 9, $p = 0.028$; net family income, $X^2 = 72.580$, df =12, $p = 0.000$.

In a recent study, Rabadán and Bernabéu [33] suggested that globalization tends to reduce the level of FN worldwide as exposure to different cultures increases [36]. The association between lower FN and higher individual openness was initially proposed by Knaapila, Silventoinen, Broms, Rose, Perola, Kaprio, and Tuorila [35]. To evaluate the association between personal openness and technological innovation in the wine sector, consumers were asked to evaluate their degree of agreement with different statements (Table 6). The results showed that according to the statistical differences reported for some items, a higher openness could be attributed to the less neophobic segments, i.e., they are more willing to travel and discover new cultures. However, in the statements covering the self-reported attachment to tradition, a clear pattern did not appear.

Table 6. Impacts of WN and FTN on statements about individual openness.

	Neophilics (n = 102)	Neophilics Anti-NT (n = 92)	Neophobics Pro-NT (n = 112)	Neophobics (n = 92)
I consider myself a traditional person (p = 0.006)	5.58 ^{ab}	6.30 ^{a}	4.93 ^{b}	5.89 ^{ab}
Religion is very important to me (p = 0.001)	4.84 ^{ab}	3.76 ^{b}	5.31 ^{a}	4.99 ^{ab}
I support the traditional values of the family (p = 0.000)	5.01 ^{b}	6.48 ^{a}	6.85 ^{a}	7.04 ^{a}
I like traveling and discovering new cultures (p = 0.000)	8.15 ^{a}	8.04 ^{a}	6.58 ^{b}	6.61 ^{b}
I believe that immigration is positive for my country (p = 0.351)	4.71	5.06	4.52	4.59

Means of the values for the consumers in each segment. Likert 9-point scale from strongly agree (1) to strongly disagree (9). Different letters in the same row mean significant differences in the consumers' opinions ($p < 0.05$).

4. Conclusions

This study shows that wine consumers show different levels of WN and FTN. This result supports previous findings stating that WN and FTN are useful to measure different aspects of FN. Beyond this, the study suggests that wine consumers with a higher income and level of education tend to be more open to trying new wines. This confirms that the idea of lower education and income leading to higher FN scores is also applicable to the wine sector. Results also state that WN reduces as product involvement increases. For this reason, neophilic consumers consume wine more frequently and consider a wider range of attributes when purchasing wine.

Regarding practical implications, obtained results encourage wine makers to develop new wines and novel production techniques, focusing on consumers with higher socioeconomic status. Within wine attributes, wine makers should pay special attention to the type of wine, the wine barrel aging, and the brand image as these are the key attributes considered by this consumer segment in the purchasing process. Wine makers should also consider that consumer segments eager to try innovative wines exist; however, not all of them are open to the same kinds of innovations, and wines including specific innovations intended for specific consumers segments would need to be developed.

Although wine is considered a traditional food product deeply linked to heritage, wine innovation is crucial to ensure increased demand, mainly in the traditional wine producing countries of Europe, and specifically in Spain, where wine consumption is significantly lower than in surrounding countries. Consumers are accustomed to continuous innovations in the food sector, and the wine sector cannot only rely on their strong connections with *terroir* and culture to survive. Moreover, severe global competition and the positive attitudes toward the innovation of consumers in emerging markets encourage these innovations.

There were some limitations to this work. First, surveys were only conducted in three locations, while the results have been extrapolated to the whole country. It should also be considered that due to the sampling method, some bias could appear. In addition, the use of different Likert-type scales may have led to confusion for some of the interviewees. Second, in market research, there may be a difference between what consumer respondents say and what their real attitudes are. Third, it should be considered that the present study has only evaluated consumer attitudes toward technological innovations (product and process), while nontechnological innovations have been neglected.

Future studies should examine the specific innovations that wine consumers are more willing to accept. Recent knowledge suggests that innovations considered too radical could face rejection even among more pro-innovation consumers. In this regard, specific studies about consumers' acceptance of nontechnological innovations, including marketing and organizational innovations, should be developed.

Funding: This research received no external funding.

Institutional Review Board Statement: The study was conducted according to the guidelines of the Declaration of Helsinki and approved by the University of Castilla-La Mancha.

Informed Consent Statement: Informed consent was obtained from all subjects involved in the study.

Data Availability Statement: Data will be available under request.

Acknowledgments: This study was partially funded by the Provincial Council of Albacete, the University of Castilla-La Mancha, and the European Regional Development Fund [2018/11744].

Conflicts of Interest: The author declares no conflict of interest.

References

1. Kühne, B.; Vanhonacker, F.; Gellynck, X.; Verbeke, W. Innovation in traditional food products in Europe: Do sector innovation activities match consumers' acceptance? *Food Qual. Prefer.* **2010**, *21*, 629–638. [CrossRef]
2. Castellini, A.; Samoggia, A. Millennial consumers' wine consumption and purchasing habits and attitude towards wine innovation. *Wine Econ. Policy* **2018**, *7*, 128–139. [CrossRef]

3. Frigon, A.; Doloreux, D.; Shearmur, R. Drivers of eco-innovation and conventional innovation in the canadian wine industry. *J. Clean. Prod.* **2020**, *275*. [CrossRef]
4. Choi, H.; Gu, C. Geospatial response for innovation in the wine industry: Knowledge creation through institutional mobility in China. *Agronomy* **2020**, *10*, 495. [CrossRef]
5. Nguyen, A.N.H.; Johnson, T.E.; Jeffery, D.W.; Danner, L.; Bastian, S.E.P. A cross-cultural examination of Australian, Chinese and Vietnamese consumers' attitudes towards a new Australian wine product containing Ganoderma lucidum extract. *Food Res. Int.* **2019**, *115*, 393–399. [CrossRef] [PubMed]
6. Ristic, R.; Johnson, T.E.; Meiselman, H.L.; Hoek, A.C.; Bastian, S.E.P. Towards development of a Wine Neophobia Scale (WNS): Measuring consumer wine neophobia using an adaptation of The Food Neophobia Scale (FNS). *Food Qual. Prefer.* **2016**, *49*, 161–167. [CrossRef]
7. Vanhonacker, F.; Kühne, B.; Gellynck, X.; Guerrero, L.; Hersleth, M.; Verbeke, W. Innovations in traditional foods: Impact on perceived traditional character and consumer acceptance. *Food Res. Int.* **2013**, *54*, 1828–1835. [CrossRef]
8. Baregheh, A.; Rowley, J.; Sambrook, S.; Davies, D. Innovation in food sector SMEs. *J. Small Bus. Enterp. Dev.* **2012**, *19*, 300–321. [CrossRef]
9. Triguero, A.; Moreno-Mondéjar, L.; Davia, M.A. Drivers of different types of eco-innovation in European SMEs. *Ecol. Econ.* **2013**, *92*, 25–33. [CrossRef]
10. OECD, O.M. The Measurement Of Scientific And Technological Activities—Oslo Manual: Guidelines For Collecting And Interpreting Innovation Data. *OECD Sci. Inf. Technol.* **2005**, *2005*, i-166.
11. Rabadán, A.; Triguero, Á.; Gonzalez-Moreno, Á. Cooperation as the secret ingredient in the recipe to foster internal technological eco-innovation in the agri-food industry. *Int. J. Environ. Res. Public Health* **2020**, *17*, 2588. [CrossRef] [PubMed]
12. Borrello, M.; Cembalo, L.; Vecchio, R. Consumers' acceptance of fungus resistant grapes: Future scenarios in sustainable winemaking. *J. Clean. Prod.* **2021**, *307*, 127318. [CrossRef]
13. Cebrián-Tarancón, C.; Sánchez-Gómez, R.; Cabrita, M.J.; García, R.; Zalacain, A.; Alonso, G.L.; Salinas, M.R. Winemaking with vine-shoots. Modulating the composition of wines by using their own resources. *Food Res. Int.* **2019**, *121*, 117–126. [CrossRef] [PubMed]
14. Sillani, S.; Miccoli, A.; Nassivera, F. Different preferences for wine communication. *Wine Econ. Policy* **2017**, *6*, 28–39. [CrossRef]
15. Cuerva, M.C.; Triguero-Cano, Á.; Córcoles, D. Drivers of green and non-green innovation: Empirical evidence in Low-Tech SMEs. *J. Clean. Prod.* **2014**, *68*, 104–113. [CrossRef]
16. Grunert, K.G. Trends in food choice and nutrition. In *Consumer Attitudes to Food Quality Products: Emphasis on Southern Europe*; Klopčič, M., Kuipers, A., Hocquette, J.-F., Eds.; Wageningen Academic Publishers: Wageningen, The Netherlands, 2013; pp. 23–30. [CrossRef]
17. Verhagen, H.; Alonso-Andicoberry, C.; Assunção, R.; Cavaliere, F.; Eneroth, H.; Hoekstra, J.; Koulouris, S.; Kouroumalis, A.; Lorenzetti, S.; Mantovani, A.; et al. Risk-benefit in food safety and nutrition—Outcome of the 2019 Parma Summer School. *Food Res. Int.* **2021**, *141*. [CrossRef] [PubMed]
18. Siegrist, M. Consumer attitudes to food innovation and technology. In *Understanding Consumers of Food Products*; Woodhead Publishing Ltd.: Cambridge, UK, 2007; pp. 236–253. [CrossRef]
19. Tuorila, H.; Hartmann, C. Consumer responses to novel and unfamiliar foods. *Curr. Opin. Food Sci.* **2020**, *33*, 1–8. [CrossRef]
20. Buisson, D. Developing new products for the consumer. In *Food Choice and the Consumer*; Chapman & Hall: Cambridge, UK, 1995; pp. 182–215.
21. Lord, J.B. New product failure and success. In *Developing New Food Products for a Changing Marketplace*; CRC Press: Boca Raton, FL, USA, 2000; pp. 55–86.
22. Rudolph, M.J. The food product development process. *Br. Food J.* **1995**, *97*, 3–11. [CrossRef]
23. Deliza, R.; Ares, G. Consumer Perception of Novel Technologies. In *Food Engineering Series*; Springer: Berlin, Germany, 2018; pp. 1–20. [CrossRef]
24. Martins, I.B.A.; Oliveira, D.; Rosenthal, A.; Ares, G.; Deliza, R. Brazilian consumer's perception of food processing technologies: A case study with fruit juice. *Food Res. Int.* **2019**, *125*, 108555. [CrossRef]
25. Henriques, A.S.; King, S.C.; Meiselman, H.L. Consumer segmentation based on food neophobia and its application to product development. *Food Qual. Prefer.* **2009**, *20*, 83–91. [CrossRef]
26. Alley, T.R. Conceptualization and measurement of human food neophobia. In *Food Neophobia*; Reilly, S., Ed.; Elsevier: Amsterdam, The Netherlands, 2018; pp. 169–192. [CrossRef]
27. Pliner, P.; Hobden, K. Development of a scale to measure the trait of food neophobia in humans. *Appetite* **1992**, *19*, 105–120. [CrossRef]
28. Cox, D.N.; Evans, G. Construction and validation of a psychometric scale to measure consumers' fears of novel food technologies: The food technology neophobia scale. *Food Qual. Prefer.* **2008**, *19*, 704–710. [CrossRef]
29. Lockshin, L.; Corsi, A.M. Consumer behaviour for wine 2.0: A review since 2003 and future directions. *Wine Econ. Policy* **2012**, *1*, 2–23. [CrossRef]
30. Niimi, J.; Danner, L.; Bastian, S.E. Wine leads us by our heart not our head: Emotions and the wine consumer. *Curr. Opin. Food Sci.* **2019**, *27*, 23–28. [CrossRef]

31. Bernabéu, R.; Prieto, A.; Díaz, M. Preference patterns for wine consumption in Spain depending on the degree of consumer ethnocentrism. *Food Qual. Prefer.* **2013**, *28*, 77–84. [CrossRef]
32. Lockshin, L.; Jarvis, W.; d'Hauteville, F.; Perrouty, J.P. Using simulations from discrete choice experiments to measure consumer sensitivity to brand, region, price, and awards in wine choice. *Food Qual. Prefer.* **2006**, *17*, 166–178. [CrossRef]
33. Rabadán, A.; Bernabéu, R. A systematic review of studies using the Food Neophobia Scale: Conclusions from thirty years of studies. *Food Qual. Prefer.* **2021**, *93*. [CrossRef]
34. Knaapila, A.; Tuorila, H.; Silventoinen, K.; Keskitalo, K.; Kallela, M.; Wessman, M.; Peltonen, L.; Cherkas, L.F.; Spector, T.D.; Perola, M. Food neophobia shows heritable variation in humans. *Physiol. Behav.* **2007**, *91*, 573–578. [CrossRef] [PubMed]
35. Knaapila, A.; Silventoinen, K.; Broms, U.; Rose, R.J.; Perola, M.; Kaprio, J.; Tuorila, H.M. Food neophobia in young adults: Genetic architecture and relation to personality, pleasantness and use frequency of foods, and body mass index-A twin study. *Behav. Genet.* **2011**, *41*, 512–521. [CrossRef]
36. Flight, I.; Leppard, P.; Cox, D.N. Food neophobia and associations with cultural diversity and socio-economic status amongst rural and urban Australian adolescents. *Appetite* **2003**, *41*, 51–59. [CrossRef]
37. Predieri, S.; Sinesio, F.; Monteleone, E.; Spinelli, S.; Cianciabella, M.; Daniele, G.M.; Dinnella, C.; Gasperi, F.; Endrizzi, I.; Torri, L.; et al. Gender, age, geographical area, food neophobia and their relationships with the adherence to the mediterranean diet: New insights from a large population cross-sectional study. *Nutrients* **2020**, *12*, 1778. [CrossRef]
38. Prescott, J.; Young, O.; O'Neill, L.; Yau, N.J.N.; Stevens, R. Motives for food choice: A comparison of consumers from Japan, Taiwan, Malaysia and New Zealand. *Food Qual. Prefer.* **2002**, *13*, 489–495. [CrossRef]
39. Salgado Beltrán, L.; Camarena Gómez, D.M.; Díaz León, J. The Mexican consumer, reluctant or receptive to new foods? *Br. Food J.* **2016**, *118*, 734–748. [CrossRef]
40. Meiselman, H.L.; King, S.C.; Gillette, M. The demographics of neophobia in a large commercial US sample. *Food Qual. Prefer.* **2010**, *21*, 893–897. [CrossRef]
41. Koivisto, U.K.; Sjödén, P.O. Reasons for rejection of food items in Swedish families with children aged 2-17. *Appetite* **1996**, *26*, 89–104. [CrossRef] [PubMed]
42. Schickenberg, B.; Van Assema, P.; Brug, J.; De Vries, N.K. Are the Dutch acquainted with and willing to try healthful food products? The role of food neophobia. *Public Health Nutr.* **2008**, *11*, 493–500. [CrossRef] [PubMed]
43. Fernández-Ruiz, V.; Claret, A.; Chaya, C. Testing a Spanish-version of the Food Neophobia Scale. *Food Qual. Prefer.* **2013**, *28*, 222–225. [CrossRef]
44. Schnettler, B.; Grunert, K.G.; Miranda-Zapata, E.; Orellana, L.; Sepúlveda, J.; Lobos, G.; Hueche, C.; Höger, Y. Testing the Abbreviated Food Technology Neophobia Scale and its relation to satisfaction with food-related life in university students. *Food Res. Int.* **2017**, *96*, 198–205. [CrossRef]
45. Schnettler, B.; Miranda-Zapata, E.; Miranda, H.; Velásquez, C.; Orellana, L.; Sepúlveda, J.; Lobos, G.; Sánchez, M.; Grunert, K.G. Psychometric analysis of the Food Technology Neophobia Scale in a Chilean sample. *Food Qual. Prefer.* **2016**, *49*, 176–182. [CrossRef]
46. Bernabéu, R.; Díaz, M.; Olivas, R.; Olmeda, M. Consumer preferences for wine applying best-worst scaling: A Spanish case study. *Br. Food J.* **2012**, *114*, 1228–1250. [CrossRef]
47. Hair, J.; Anderson, R.; Tatham, R.; Black, W. *Análisis Multivariante*; Prentice Hall: Madrid, Spain, 2007.
48. Deegan, K.C.; Palmujoki, I.; Isotalo, J.; Tuorila, H. Effective communication of novelty: The case of ripened cheese. *Food Qual. Prefer.* **2015**, *40*, 68–76. [CrossRef]
49. Verbeke, W.; Marcu, A.; Rutsaert, P.; Gaspar, R.; Seibt, B.; Fletcher, D.; Barnett, J. 'Would you eat cultured meat?': Consumers' reactions and attitude formation in Belgium, Portugal and the United Kingdom. *Meat Sci.* **2015**, *102*, 49–58. [CrossRef] [PubMed]
50. Verbeke, W. Profiling consumers who are ready to adopt insects as a meat substitute in a Western society. *Food Qual. Prefer.* **2015**, *39*, 147–155. [CrossRef]
51. Gere, A.; Székely, G.; Kovács, S.; Kókai, Z.; Sipos, L. Readiness to adopt insects in Hungary: A case study. *Food Qual. Prefer.* **2017**, *59*, 81–86. [CrossRef]
52. Rabadán, A.; Martínez-Carrasco, L.; Brugarolas, M.; Bernabéu, R. Perceptions of geographical indication labels as quality indicators inside and outside the labels' area of influence: The case of spring fruits. *Renew. Agric. Food Syst.* **2021**. [CrossRef]
53. Casali, G.L.; Perano, M.; Presenza, A.; Abbate, T. Does innovation propensity influence wineries' distribution channel decisions? *Int. J. Wine Bus. Res.* **2018**, *30*, 446–462. [CrossRef]
54. Meiselman, H.L. A longitudal study of food habits, attitudes and human performance. *Appetite* **1995**, *24*, 193.
55. Costa, A.; Silva, C.; Oliveira, A. Food neophobia and its association with food preferences and dietary intake of adults. *Nutr. Diet.* **2019**. [CrossRef] [PubMed]
56. Nordin, S.; Broman, D.A.; Garvill, J.; Nyroos, M. Gender differences in factors affecting rejection of food in healthy young Swedish adults. *Appetite* **2004**, *43*, 295–301. [CrossRef]
57. Dovey, T.M.; Staples, P.A.; Gibson, E.L.; Halford, J.C.G. Food neophobia and 'picky/fussy' eating in children: A review. *Appetite* **2008**, *50*, 181–193. [CrossRef] [PubMed]

foods

Article

Elaboration of Gluten-Free Cookies with Defatted Seed Flours: Effects on Technological, Nutritional, and Consumer Aspects

Elena Martínez [1], Rita García-Martínez [1], Manuel Álvarez-Ortí [1], Adrián Rabadán [1], Arturo Pardo-Giménez [2] and José E. Pardo [1,*]

[1] E.T.S.I. Agrónomos y de Montes (UCLM), Campus Universitario s/n, 02071 Albacete, Spain; elena.martinez44@alu.uclm.es (E.M.); rita.garcia@alu.uclm.es (R.G.-M.); manuel.alvarez@uclm.es (M.Á.-O.); adrian.rabadan@uclm.es (A.R.)

[2] Centro de Investigación, Experimentación y Servicios del Champiñón (CIES), C/Peñicas s/n, Apdo. 63, 16220-Quintanar del Rey, 160220 Cuenca, Spain; apardo.cies@dipucuenca.es

* Correspondence: jose.pgonzalez@uclm.es

Abstract: Cookies, which form the largest category of bakery snacks, are considered a good vehicle to introduce nutrients into the diet. In this study, to increase the nutritional value of traditional commercial cookies, wheat flour was substituted with defatted flours made from flax, sesame, chia, and poppy, which are byproducts of the oil extraction industry. The differences in the technological properties, nutritional composition, and consumer acceptance of the reformulated cookies were evaluated. The results show that the wheat cookies used as the control showed a more elastic behavior than the cookies elaborated with defatted seed flours, which showed a greater tendency to crumble. The use of defatted seed flours yielded cookies with a higher content of protein and fiber, and a lower content in carbohydrates than the wheat cookies. Consumer evaluations for the sesame and flax cookies were similar to those for the traditional wheat cookies, with positive assessments on all of the parameters evaluated. On the other hand, the cookies elaborated using chia and poppy flours received the least positive evaluations from consumers. Thus, the use of some defatted seed flours, mainly flax and sesame, is proposed as an interesting alternative to produce health-promoting cookies in order to cover the current demand for gluten-free products.

Keywords: food innovation; chia; flax; proximate composition; sesame; poppy

Citation: Martínez, E.; García-Martínez, R.; Álvarez-Ortí, M.; Rabadán, A.; Pardo-Giménez, A.; Pardo, J.E. Elaboration of Gluten-Free Cookies with Defatted Seed Flours: Effects on Technological, Nutritional, and Consumer Aspects. *Foods* **2021**, *10*, 1213. https://doi.org/10.3390/foods10061213

Academic Editor: Grant Campbell

Received: 13 May 2021
Accepted: 26 May 2021
Published: 27 May 2021

1. Introduction

Cookies are baked products containing three major ingredients: flour, fat, and sugar. They form the largest category of bakery snacks because of their low cost, good taste, texture (crispness), and storability, and are considered an effective vehicle for nutrient supply to consumers. Cookies are usually developed with wheat flour because it forms unique visco-elastic dough when mixed with water, due to the presence of gluten [1]. However, in specific individuals, gluten may create autoimmune reactions. For this reason, the demand for gluten-free products is increasing, leading to a considerable growth in the gluten-free food market [2]. Even among gluten-tolerant individuals, a reduction in the popularity of products with gluten has been observed in recent years [3]. Thus, it would be interesting to substitute wheat for other gluten-free flours that enhance the nutritional quality of the final product [4,5].

However, the nutritional profile of gluten-free bakery products available in the market is often questioned [6,7]. It is thus advisable to replace wheat flour with other flours that provide nutrients and other healthy compounds, and to thus produce cookies that may contribute to the design of a healthy diet. In addition, the healthy properties of snacks are currently a major concern [8,9]. For this purpose, the use of seed defatted flours, which are byproducts of the oil extraction industry, may be appropriate. Several seeds that are commercially available for human consumption as food supplements are considered functional foods because of their beneficial effects on health. This category includes flax, sesame, chia, and poppy seeds. These contain high levels of oil (higher than 20%), which can be extracted by pressure systems, resulting in an oil extraction industry that produces high quality oils rich in polyunsaturated fatty acids. One byproduct of this industry is the press cake, which, once ground, yields the defatted flour. Although these defatted flours may be used for other purposes [10,11], they are generally considered as waste, with no further use. Thus, the incorporation of defatted seed flours into the formulation of cookies may be useful to increase the added value of this byproduct [12].

In addition, defatted seed flours may contain valuable compounds to improve the nutritional properties of cookies. They show higher levels of proteins and total dietary fiber, but a lower content of fat than the seeds [13,14]. However, the remaining fat content has the benefit of seed oils, where polyunsaturated fatty acids are predominant [15]. Additionally, these seeds contain a high proportion of bioactive compounds, such as polyphenolic compounds, with antioxidant properties [16].

In this work, we evaluated the physical, nutritional, and sensory behavior of cookies elaborated with defatted flour from chia, flax, sesame, and poppy seeds, in order to consider the use of this valuable byproduct from the oil industry in the food chain.

2. Materials and Methods

2.1. Raw Materials

All of the seeds used in this study were acquired in local supermarkets. To prepare the defatted flours, seeds from flax (*Linum usitatissimum*), sesame (*Sesamum indicum*), chia (*Salvia hispanica*), and poppy (*Papaver somniferum*) were subjected to oil extraction with a screw press (Komet Oil Press CA59G, IBG Monforts Oekotec GmbH and Co. KG, Mönchengladbach, Germany) at 49 rpm and 75 °C [17].

2.2. Cookies Preparation

The recipe adopted for cookie preparation followed the process proposed by Jan et al. (2018) [4], with slight modifications. First, 25 g brown sugar and 25 g of refined sunflower oil as a fat source were mixed to obtain a creamy mixture. The rest of the ingredients were then added to form the dough, as follows: flour 50 g, sodium bicarbonate 1 g, salt 0.5 g, skimmed powder milk 2.5 g, and water 8 mL. The dough was rolled out with a rolling pin to a uniform thickness of 0.6 cm and cut to form 5 cm diameter round cookies. These were then baked at 200 °C for 10 min. The cookies were subsequently cooled before performing the analysis.

For each type of cookie, the wheat flour as well as defatted flax, sesame, chia, and poppy flours were used separately.

2.3. Physical Measurements

We measured the diameter and the height of 10 baked cookies of each type, using a digital caliper. The spreading factor was calculated by dividing the diameter by the height.

The color was measured by reflection in five random points on the surface of the cookies with a Minolta CR-200 colorimeter (Minolta Camera Co. Ltd., Osaka, Japan), using D65 as the illuminant. The tristimulus values were used to calculate the CIELab chromatic coordinates, as follows: L^* (brightness), a^* (red-green component), and b^* (yellow-blue component).

To measure the texture of the cookies, five cookies of each type were cut perpendicularly with a Warner Bratzler blade with a rectangular slot blade at a constant velocity of 2 mm/s, using a TA-XT Plus texture analyzer (Stable Micro Systems, Godalming, UK). We recorded the maximum force needed to cut the cookie and the deformation before breaking.

2.4. Proximate Composition

The main nutritional components of the cookies were measured following Rabadán, et al. [18]. Briefly, the protein content was calculated by multiplying the total nitrogen content, obtained by the Kjeldahl method, by a conversion factor of 6.25. To determine the ash content, the flours were ashed at 550 °C to a constant weight. Crude fat (ether extract) was estimated gravimetrically using the filter bag technique after petroleum ether extraction of the dried sample in an Ankom XT10 extraction system. To determine the content of crude fiber, we applied the Weende technique, adapted to the filter bag technique. This method determines the organic residue remaining after digestion with solutions of sulfuric acid and sodium hydroxide, using an Ankom 220 fiber analyzer. The total carbohydrate content was calculated by subtracting the sum of the crude protein, total fat, water, and ash from the total weight of the flour [19]. The available carbohydrate content (nitrogen-free) was calculated by subtracting the crude fiber from the total carbohydrate content [20]. The energy value of the cookies was estimated from the relative content of the protein (N $\times$ 6.25), fat, and carbohydrates, using the Atwater general factors of 4.0, 9.0, and 4.0 kcal g-1 for each component, respectively. The proximate analysis was performed in triplicate for each cookie formulation.

2.5. Consumer Preferences

To measure consumers' preferences for the cookies, 106 participants were selected among staff and students at the University of Castilla-La Mancha. Only regular consumers of cookies were involved in the study. With that purpose in mind, only individuals who reported having eaten cookies at least once in the last month were selected for the study. Consumers were asked to score on an 11-point scale how much they liked the recipe, from least (0, I do not like it at all) to most (10, I like it very much). Each consumer spent about 15 min evaluating the cookies. The considered parameters were color, odor, taste, appearance, texture (crunchiness), and overall acceptability of the cookie.

2.6. Statistical Analysis

Statistical differences were estimated using an ANOVA test at a 5% level ($p \leq 0.05$) of significance. All of the statistical analyses were carried out using the SPSS program, release 23.0 for Windows.

3. Results and Discussion

3.1. Physical Parameters

The spread factor, measured by dividing the diameter of the cookies by their height, is an important parameter to estimate the behavior of the dough during baking. A higher spread factor and larger diameter are considered as crucial quality characteristics for cookies [21]. Although it has been proposed that the gluten, sugar, or fiber content may influence the spread factor [22], in this case, the behavior of cookies was unclear. The spread factor varied from 8.97 in the flax cookies to 11.10 in the poppy cookies, while the cookies made with wheat flour, with gluten, showed intermediate values (Table 1). Significant differences were found in the values between the cookies made with sesame and poppy seed flour. It has been proposed that gluten forms a web during the baking of cookies, which increases the viscosity and stops the flow of cookie dough, leading to lower cookie diameters [23]. However, the diameter of the wheat cookies showed no significant differences with those obtained in the rest of the formulations. This could be explained by the existence of other proteins that may also affect the viscosity of the dough. Regarding the cookies made with defatted seed flours, no clear correlations between spread factor and the rest of nutritional parameters measured were found. Regarding the diameter, the sesame cookies showed significant differences compared with the chia and flax cookies, which showed the lowest values.

Table 1. Mean values for size parameters (diameter and height) and spread factor for all of the cookies.

	Diameter (cm)	Height (cm)	Spread Factor
Wheat	5.34 ± 0.27 [ab]	0.57 ± 0.03 [ab]	9.20 ± 0.13 [b]
Flax	5.22 ± 0.34 [b]	0.59 ± 0.02 [a]	8.97 ± 0.97 [b]
Sesame	5.62 ± 0.22 [a]	0.54 ± 0.01 [abc]	10.20 ± 0.16 [ab]
Chia	5.11 ± 0.21 [b]	0.51 ± 0.02 [bc]	9.78 ± 0.37 [ab]
Poppy	5.31 ± 0.21 [ab]	0.49 ± 0.06 [c]	11.10 ± 1.38 [a]

Numbers are means of multiple measurements. Different letters in the same column indicate significant differences ($p < 0.05$).

Color was measured according to CIE L*a*b* parameters, where L* represents lightness, a* the value in the red-green axis, and b* the value in the yellow-blue axis. The use of different ingredients has significant effects on the color of the cookies [24]. Surface color is considered an important indicator of the degree of baking, and may play an important role in consumer acceptance. The expected color for cookies is golden brown for the surface and creamish white for the crumb [25]. However, in this case, the color of the cookies was greatly affected by the color of the flour. Chia and poppy flours were darker, resulting in cookie colors with lower L* values. Although baking tends to decrease the lightness of cookies, reducing the difference between them, it is still possible to appreciate significant differences in this parameter (Figure 1A). When the color of the cookies was measured, only those elaborated with flax seed defatted flour showed a similar color to the ones elaborated with wheat flour (Figure 1B). The cookies made from chia and poppy seed flours formed another group with lower values of a* and b*, while the cookies made with sesame flour showed intermediate values. Similar results have been observed when other gluten-free flours are used to make cookies [26]. The lower values of L*, a*, and b* in the cookies elaborated with defatted seed flour indicate less attractive colors for consumers [25]; although, in the case of flax and sesame, the differences with the wheat cookies were smaller.

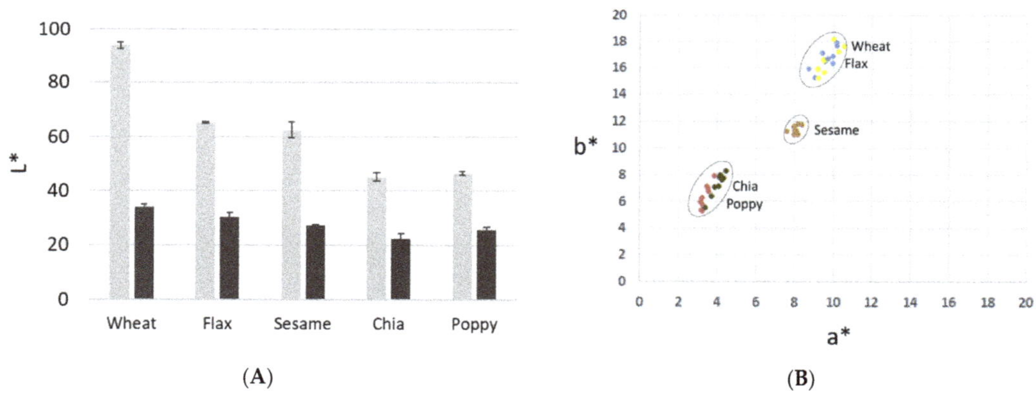

Figure 1. (**A**) Values of brightness (L*) of flour (light grey bars) and cookies (dark grey bars). (**B**) Values of CIELab color coordinates a* (red-green) and b* (yellow-blue) of the cookies elaborated with the different defatted seed flours.

Regarding texture, we recorded the maximum force needed to break the cookies and the deformation until the cookies broke (Figure 2). Different patterns were observed for this parameter. The wheat cookies, used as the control, showed a more elastic behavior, represented by the lower slope in Figure 2. This means that the cookies were deformed to a greater degree when force was applied until breaking occurred. The other cookies elaborated with defatted seed flours showed a more fragile behavior, as the deformation was lower when the force was applied. Regarding the maximum force needed to break the cookies, those elaborated with flax flour were the hardest (53.72 ± 11.01 N), followed by chia cookies (39.42 ± 11.42 N), leading to crispier cookies. The cookies elaborated with defatted flours from sesame and poppy showed a similar breaking force to wheat cookies, although deformation until breaking was lower, indicating a more fragile behavior.

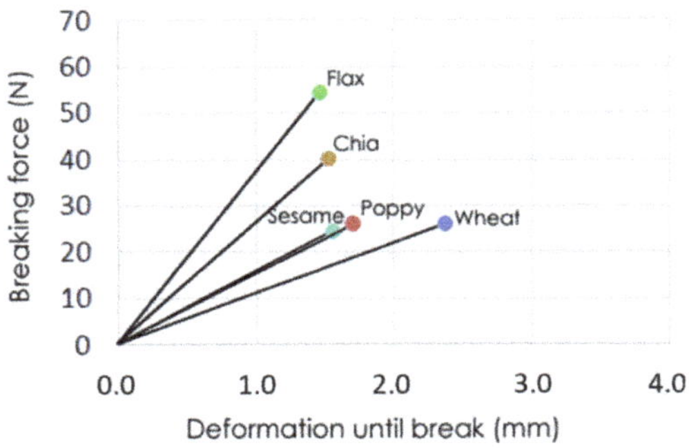

Figure 2. Maximum force and deformation until breaking for cookies elaborated with defatted seed flours. Points are the average value of five repetitions.

3.2. Proximate Composition

Table 2 shows the proximate composition of the cookies elaborated with the different seed flours. Humidity was low in all of the cookies, and was lower than 2% in all cases. The reduced values of humidity are adequate to ensure a long shelf-life, making the cookies a long-life food if adequately packaged.

Table 2. Proximate composition of cookies elaborated with different defatted seed flours.

	Wheat	Flax	Sesame	Chia	Poppy
Humidity (%)	1.6	1.0	1.4	0.8	1.8
Nitrogen (%)	0.96	2.47	2.68	2.55	2.74
Protein (%)	6.00	15.44	16.75	15.94	17.13
Ashes (%)	1.34	3.38	4.27	4.13	6.04
Fiber (%)	0.75	4.03	2.79	12.41	6.96
Fat (%)	4.29	5.93	14.11	5.22	7.07
Total carbohydrates (%)	88.37	75.25	64.87	74.11	69.77
Available carbohydrates (%)	87.62	71.22	62.08	62.30	62.81
Energy value (kcal/100 g)	416	416	453	410	411

Numbers are means of three independent measurements.

One of the main advantages of seed defatted flours is their high content of proteins [14], which is an interesting source to enhance the nutritional characteristics of baked products. The cookies elaborated with defatted seed flours showed a higher protein content than the wheat cookies. The highest protein content was observed in poppy cookies (17.13%), although no significant differences in flax, sesame, and chia cookies were revealed. In all of the cases, the cookies yielded more than twice the protein content of wheat cookies. Previous studies have reported the addition of defatted sesame seed as a useful addition for increasing the protein content of cookies [27]. Although the main use of these seeds is as oil, because of the elevated proportion of polyunsaturated fatty acids, the benefits of their protein fraction in the defatted flours has also been evaluated [28].

In addition, the fiber content was also higher in all of the cookies elaborated with seed defatted flours. Fiber intake remains low in Western societies, despite the health benefits attributed to it, which are related to metabolic parameters, microbiome composition, and metabolite production [29]. Thus, the fortification of foods with fiber is a major area of interest in the food industry, and is an opportunity for food reformulation [30]. Fiber is an important component of the studied seeds as it represents, for example, 24.65% of the total composition in poppy seeds [31], 27.30% in flax seeds [32], and 34.4% in chia seeds [32]. In this sense, the cookies elaborated with defatted chia flour showed the highest fiber content (12.41%), followed by poppy, sesame, and flax. All of the reformulated cookies showed much higher values than those reported for the wheat cookies (Table 2).

Another important parameter in the proximate analysis is fat content. In this sense, seed flour cookies showed an increase in fat content due to the fat remaining in the flour after oil extraction. When pressure systems are used to extract oil, generally about 15–20% of fat remains in the defatted flour, depending on the oil extraction method [33]. In any event, seed oils show positive characteristics as they are rich in polyunsaturated fatty acids, and this oil can also be considered a healthy source of fat [31]. They are especially rich in α-linolenic acid, a fatty acid the body cannot synthesize and that is the biological precursor to eicosapentaenoic acid and docosahexaenoic acid [34]. Sesame cookies were notable for their fat content (14.11%), because of the lower yield in the oil extraction process and the consequent higher fat content in the defatted flour.

The energy values were similar in all of the cookies, except those made with sesame, due to their higher fat content. The higher protein and fat content in seed flour cookies was counteracted by a lower total carbohydrate content. To reduce the energy value of the sesame cookies, the oil extraction process should be optimized to obtain flour with a reduced fat content [33].

Regarding these data, defatted flours from seeds represent an interesting ingredient to fortify cookies from a nutritional point of view, as they increase the protein and fiber content and reduce carbohydrates. The proximate analysis results suggest that all of the reformulated cookies showed a better nutritional quality than the traditional wheat cookies [35].

3.3. Consumer Evaluation

Regarding the consumer evaluation of cookies, each participant was asked to indicate their preferences on several parameters, namely: color, odor, taste, appearance, texture, and overall acceptance. The data are shown in Figure 3. The color of the cookies made with chia and poppy seed flours was darker than the rest of cookies [24]. As previously reported, consumers prefer golden brown cookies [25], and, consequently, these dark cookies were scored lower by our consumers. On the other hand, the cookies made with flax and sesame flours had positive evaluations, although there were significant differences with the scores obtained by the wheat cookies. Drawing on previous studies, in order to obtain similar values for color in reformulated cookies to those obtained for traditional cookies, traditional flours should only be partially replaced [25]. A significant proportion of consumers are reluctant to try novel foods that differ, in color, for example, to those they usually eat [36].

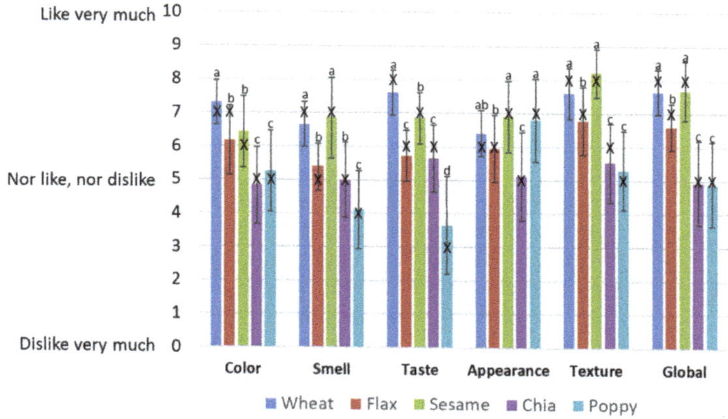

Figure 3. Sensory properties of cookies elaborated with defatted seed flours. Different letters in the same column indicate significant differences ($p < 0.05$).

Regarding odor, again the chia and poppy cookies scored the lowest. In this case, poppy cookies showed values lower than 5, indicating that the odor of these cookies was unpleasant for consumers. Being an odor that consumers are unused to, they consider it a negative attribute. A similar behavior was found when taste was considered, with the poppy cookies scoring lowest, with values below 5. The rest of cookies showed values over 5, indicating that consumers liked them. The sesame cookies showed a median value of 7, the highest of the seed defatted flour cookies, although the highest scores were obtained by the wheat cookies used as the controls. The high values reported for sesame compared with the other seeds could be attributed to the higher fat content. Higher fat content tends to increase consumer preference for various food products, compared with the low-fat versions [37,38].

Regarding texture, cookies made with chia and poppy flours showed a high tendency to crumble, again obtaining a low consumer evaluation. The other three cookies (wheat, flax, and sesame) showed similar crunchy characteristics, leading to higher scores from the consumers, with no significant differences. The global acceptability of sesame cookies showed no significant differences from those of wheat, meaning that consumers could easily change from traditional wheat cookies to 100% defatted flour sesame cookies. Previously, other studies analyzing the overall acceptability of traditional and partially reformulated sesame cookies have reported a similar acceptability in both types [27].

Flax flour cookies also showed positive evaluations in all of the categories tested. On the other hand, the cookies made with chia and poppy defatted flour showed the lowest values regarding sensory evaluation. In this case, other ingredients should be included in the cookie recipe, for example other fat sources [38], to provide better sensory properties that fit consumers' preferences. These flours could also be mixed, in a low percentage, with wheat flour or other flours to improve acceptance [39].

4. Conclusions

Here, cookies were made with seed defatted flour (flax, sesame, chia, and poppy), a byproduct of the oil extraction industry. From a physical point of view, all the cookies made showed a similar behavior when baked, with no clear differences regarding spread factor. The color of the cookies was greatly influenced by the color of the flour, resulting in darker cookies when chia and poppy flours were used. Regarding texture, seed defatted flour cookies showed a higher tendency to crumble, while wheat cookies had a more elastic behavior.

The use of flax, sesame, chia, and poppy defatted flours improves the nutritional properties of cookies, with an increase in protein, fiber, and fat content, but with a decrease in total carbohydrates. From a sensory point of view, the cookies elaborated with flax and sesame defatted flours showed similar values to those of the wheat cookies used as a control on all of the parameters evaluated (color, odor, taste, texture, appearance, and global acceptability). Chia and poppy cookies showed lower values, especially poppy cookies, which yielded values lower than 5 for odor and taste.

All these data suggest that seed defatted flours are an interesting ingredient for improving the nutritional characteristics of cookies, although in the case of chia and poppy, other ingredients, mainly other fats, should be included to improve sensory attributes.

Author Contributions: Conceptualization, M.Á.-O. and J.E.P.; methodology, E.M., R.G.-M. and A.P.-G.; validation, M.Á.-O. and J.E.P.; formal analysis, A.P.-G. and A.R.; investigation, E.M. and R.G.-M.; data curation, M.Á.-O.; writing—original draft preparation, M.Á.-O.; writing—review and editing, A.R., A.P.-G. and J.E.P.; visualization, M.Á.-O.; funding acquisition, J.E.P. All authors have read and agreed to the published version of the manuscript.

Funding: This research was funded by Castilla-La Mancha Regional Government (FEDERSB-PLY/19/180501/000047).

Institutional Review Board Statement: Not applicable.

Informed Consent Statement: Not applicable.

Data Availability Statement: Data available on request.

Conflicts of Interest: The authors declare no conflict of interest.

References

1. Rosentrater, K.A.; Evers, A. *Kent's Technology of Cereals: An Introduction for Students of Food Science and Agriculture*; Woodhead Publishing: Cambridge, UK, 2017.
2. Facts, P. *Gluten-Free Foods in the U.S.*, 6th ed.; Packaged Facts: Rockville, MD, USA, 2016.
3. De Magistris, T.; Belarbi, H.; Hellali, W. Examining Non-Celiac Consumers of Gluten-Free Products: An Empirical Evidence in Spain. In *Celiac Disease and Non-Celiac Gluten Sensitivity*; Rodrigo, L., Ed.; Headquarters IntechOpen Limited: London, UK, 2017.
4. Jan, K.N.; Panesar, P.S.; Singh, S. Optimization of antioxidant activity, textural and sensory characteristics of gluten-free cookies made from whole indian quinoa flour. *LWT* **2018**, *93*, 573–582. [CrossRef]
5. Rao, B.D.; Kulkarni, D.B.; Kavitha, C. Study on evaluation of starch, dietary fiber and mineral composition of cookies developed from 12 sorghum cultivars. *Food Chem.* **2018**, *238*, 82–86. [CrossRef]
6. Arslan, M.; Rakha, A.; Xiaobo, Z.; Mahmood, M.A. Complimenting gluten free bakery products with dietary fiber: Opportunities and constraints. *Trends Food Sci. Technol.* **2019**, *83*, 194–202. [CrossRef]
7. Melini, V.; Melini, F. Gluten-Free Diet: Gaps and Needs for a Healthier Diet. *Nutrients* **2019**, *11*, 170. [CrossRef]
8. Damen, F.W.M.; Hofstede, G.J.; Steenbekkers, B.L.P.A.; Vitaglione, P.; Pellegrini, N.; Fogliano, V.; Luning, P.A. Values and value conflicts in snack providing of Dutch, Polish, Indonesian and Italian mothers. *Food Res. Int.* **2019**, *115*, 554–561. [CrossRef]
9. Damen, F.W.M.; Luning, P.A.; Fogliano, V.; Steenbekkers, B.L.P.A. What influences mothers' snack choices for their children aged 2–7? *Food Qual. Prefer.* **2019**, *74*, 10–20. [CrossRef]
10. Pardo-Giménez, A.; Catalán, L.; Carrasco, J.; Álvarez-Ortí, M.; Zied, D.; Pardo, J. Effect of supplementing crop substrate with defatted pistachio meal on *Agaricus bisporus* and *Pleurotus ostreatus* production. *J. Sci. Food Agric.* **2016**, *96*, 3838–3845. [CrossRef] [PubMed]
11. Pardo-Giménez, A.; Zied, D.C.; Álvarez-Ortí, M.; Rubio, M.; Pardo, J.E. Effect of supplementing compost with grapeseed meal on *Agaricus bisporus* production. *J. Sci. Food Agric.* **2012**, *92*, 1665–1671. [CrossRef]
12. Şahin, S.; Elhussein, E.A.A. Assessment of sesame (*Sesamum indicum* L.) cake as a source of high-added value substances: From waste to health. *Phytochem. Rev.* **2018**, *17*, 691–700. [CrossRef]
13. Fernández-López, J.; Lucas-González, R.; Viuda-Martos, M.; Sayas-Barberá, E.; Pérez-Alvarez, J.A. Chia Oil Extraction Coproduct as a Potential New Ingredient for the Food Industry: Chemical, Physicochemical, Techno-Functional and Antioxidant Properties. *Plant Foods Human Nutr.* **2018**, *73*, 130–136. [CrossRef]
14. Prakash, K.; Naik, S.N.; Vadivel, D.; Hariprasad, P.; Gandhi, D.; Saravanadevi, S. Utilization of defatted sesame cake in enhancing the nutritional and functional characteristics of biscuits. *J. Food Process. Preserv.* **2018**, *42*, e13751. [CrossRef]
15. Ewerlin, M.; Bortolatto, R.; Torquato, A.; Steinmacher, N.; Rodrigues, A. Proximate Composition and Fatty Acids from Partially Defatted Chia (*Salvia hispanica*) Flour. *Rev. Virtual Química* **2018**, *10*, 13–20. [CrossRef]
16. Pellegrini, M.; Lucas-Gonzalez, R.; Sayas-Barberá, E.; Fernández-López, J.; Pérez-Álvarez, J.A.; Viuda-Martos, M. Bioaccessibility of Phenolic Compounds and Antioxidant Capacity of Chia (*Salvia hispanica* L.) Seeds. *Plant Foods Hum. Nutr.* **2018**, *73*, 47–53. [CrossRef] [PubMed]
17. Rabadán, A.; Pardo, J.E.; Gómez, R.; Álvarez-Ortí, M. Influence of temperature in the extraction of nut oils by means of screw pressing. *LWT* **2018**, *93*, 354–361. [CrossRef]
18. Rabadán, A.; Álvarez-Ortí, M.; Gómez, R.; Pardo-Giménez, A.; Pardo, J.E. Characterization of pistachio oils and defatted flours regarding cultivar and geographic origin. *J. Food Compos. Anal.* **2018**, *71*, 56–64. [CrossRef]
19. Sullivan, D.M. Proximate and mineral analysis. In *Methods of Analysis for Nutrition Labeling*; Sullivan, D.M., Carpenter, D.E., Eds.; AOAC International: Arlington, VA, USA, 1993.
20. González, J.; Alvira, P.; González, G. La cascarilla de arroz en la alimentación animal. II. Composición químico-bromatológica. *Rev. Agroquímica Tecnol. Aliment.* **1987**, *27*, 139–149.
21. Panghal, A.; Chhikara, N.; Khatkar, B.S. Effect of processing parameters and principal ingredients on quality of sugar snap cookies: A response surface approach. *J. Food Sci. Technol.* **2018**, *55*, 3127–3134. [CrossRef]
22. Yamsaengsung, R.; Berghofer, E.; Schoenlechner, R. Physical properties and sensory acceptability of cookies made from chickpea addition to white wheat or whole wheat flour compared to gluten-free amaranth or buckwheat flour. *Int. J. Food Sci. Technol.* **2012**, *47*, 2221–2227. [CrossRef]
23. Ganorkar, P.; Jain, R. Effect of flaxseed incorporation on physical, sensorial, textural and chemical attributes of cookies. *Int. Food Res. J.* **2014**, *21*, 1515–1521.
24. Aksoylu, Z.; Çağindi, O.; Köse, E. Effects of blueberry, grape seed powder and poppy seed incorporation on physicochemical and sensory properties of biscuit. *J. Food Qual.* **2015**, *38*, 164–174. [CrossRef]
25. Awolu, O.O.; Sudha, L.M.; Manohar, B. Influence of defatted mango kernel seed flour addition on the rheological characteristics and cookie making quality of wheat flour. *Food Sci. Nutr.* **2018**, *6*, 2363–2373. [CrossRef]
26. Sulieman, A.A.; Zhu, K.-X.; Peng, W.; Hassan, H.A.; Mahdi, A.A.; Zhou, H.-M. Influence of fermented and unfermented *Agaricus bisporus* polysaccharide flours on the antioxidant and structural properties of composite gluten-free cookies. *LWT* **2019**, *101*, 835–846. [CrossRef]

27. Sodipo, M.A.; Oluwamukomi, M.O.; Oderinde, Z.A.; Awolu, O.O. Nutritional Evaluation of Unripe Plantain, Moringa Seed and Defatted Sesame Seed Cookies. *Int. J. Food Stud.* **2021**, *10*, SI72–SI81. [CrossRef]
28. Cotabarren, J.; Rosso, A.M.; Tellechea, M.; García-Pardo, J.; Rivera, J.L.; Obregón, W.D.; Parisi, M.G. Adding value to the chia (*Salvia hispanica* L.) expeller: Production of bioactive peptides with antioxidant properties by enzymatic hydrolysis with Papain. *Food Chem.* **2019**, *274*, 848–856. [CrossRef] [PubMed]
29. Farooqui, A.A. Importance and Roles of Fiber in the Diet. In *High Calorie Diet and the Human Brain*; Springer: Berlin, Germany, 2015; pp. 193–218.
30. Mudgil, D.; Barak, S.; Khatkar, B.S. Cookie texture, spread ratio and sensory acceptability of cookies as a function of soluble dietary fiber, baking time and different water levels. *LWT* **2017**, *80*, 537–542. [CrossRef]
31. Musa Özcan, M.; Atalay, Ç. Determination of seed and oil properties of some poppy (*Papaver somniferum* L.) varieties. *Grasas Aceites* **2006**, *57*, 169–174. [CrossRef]
32. Alasalvar, C.; Chang, S.K.; Bolling, B.; Oh, W.Y.; Shahidi, F. Specialty seeds: Nutrients, bioactives, bioavailability, and health benefits: A comprehensive review. *Compr. Rev. Food Sci. Food Saf.* **2021**, *20*, 2382–2427. [CrossRef] [PubMed]
33. Rabadán, A.; Álvarez-Ortí, M.; Gómez, R.; Alvarruiz, A.; Pardo, J.E. Optimization of pistachio oil extraction regarding processing parameters of screw and hydraulic presses. *LWT Food Sci. Technol.* **2017**, *83*, 79–85. [CrossRef]
34. Parker, J.; Schellenberger, A.N.; Roe, A.L.; Oketch-Rabah, H.; Calderón, A.I. Therapeutic Perspectives on Chia Seed and Its Oil: A Review. *Planta Med.* **2018**, *84*, 606–612. [CrossRef]
35. Van Dam, R.M.; Seidell, J.C. Carbohydrate intake and obesity. *Eur. J. Clin. Nutr.* **2007**, *61*, S75–S99. [CrossRef]
36. Rabadán, A.; Bernabéu, R. A systematic review of studies using the Food Neophobia Scale: Conclusions from thirty years of studies. *Food Qual. Prefer.* **2021**, *93*, 104241. [CrossRef]
37. Olivares, A.; Navarro, J.L.; Salvador, A.; Flores, M. Sensory acceptability of slow fermented sausages based on fat content and ripening time. *Meat Sci.* **2010**, *86*, 251–257. [CrossRef] [PubMed]
38. Li, Z.; Marshall, R.; Heymann, H.; Fernando, L. Effect of Milk Fat Content on Flavor Perception of Vanilla Ice Cream1. *J. Dairy Sci.* **1997**, *80*, 3133–3141. [CrossRef]
39. Atuonwu, A.C.; Akobundu, E.N.T. Nutritional and sensory quality of cookies supplemented with defatted pumpkin (*Cucurbita pepo*) seed flour. *Pak. J. Nutr.* **2010**, *9*, 672–677. [CrossRef]

foods

MDPI

Article

Potential Use of Vacuum Impregnation and High-Pressure Homogenization to Obtain Functional Products from Lulo Fruit (*Solanum quitoense* Lam.)

Leidy Indira Hinestroza-Córdoba [1,2], Cristina Barrera [2], Lucía Seguí [2] and Noelia Betoret [2,*]

[1] Grupo de Valoración y Aprovechamiento de la Biodiversidad, Universidad Tecnológica del Chocó, AA.292, Calle 22 No. 18B-10, 270002 Quibdó, Colombia; leihicor@doctor.upv.es

[2] Institute of Food Engineering for Development, Universitat Politècnica de València, Camino de Vera s/n, 46022 València, Spain; mcbarpu@tal.upv.es (C.B.); lusegil@upvnet.upv.es (L.S.)

* Correspondence: noebeval@tal.upv.es

Abstract: Lulo (*Solanum quitoense* Lam.) is a Colombian fruit that is mostly used in the preparation of homemade juice as well as natural remedy for hypertension. The aim of this study was to determine physicochemical and antioxidant properties (antioxidant capacity, total phenols, flavonoids and spermidine content, and polyphenolic compounds profile by liquid chromatography—mass spectrometry (LC-MS)) of the lulo fruit and its juice. Additionally, vacuum impregnation (VI) properties of the fruit and the effect of high homogenization pressure (50, 100, and 150 MPa) on the juice properties were studied. The results revealed a good availability and impregnation capacity of the pores in fruits with similar maturity index. The main differences observed between the juice and fruit derive from removing solids and bioactive components in the filtering operation. However, the effect of high-pressure homogenization (HPH) on particle size and bioactive compounds increases the antiradical capacity of the juice and the diversity in polyphenolics when increasing the homogenization pressure.

Keywords: LC-mass spectrometry; antioxidant capacity; vacuum impregnation; polyphenolic profile; lulo fruit juice; spermidine

Citation: Hinestroza-Córdoba, L.I.; Barrera, C.; Seguí, L.; Betoret, N. Potential Use of Vacuum Impregnation and High-Pressure Homogenization to Obtain Functional Products from Lulo Fruit (*Solanum quitoense* Lam.). *Foods* **2021**, *10*, 817. https://doi.org/10.3390/foods10040817

Academic Editors: Adrián Rabadán and Rodolfo Bernabéu

Received: 10 March 2021
Accepted: 6 April 2021
Published: 9 April 2021

1. Introduction

Having great fruit diversity, Colombia is worldwide considered one of the main producers of tropical fresh fruit with advantageous possibilities for the development of healthy food products from regional wild species [1].

One of Colombia's territorial departments with great natural wealth is Chocó, which is characterized by its mega diverse and complex ecosystems with high primary production and little agrifood system transformation. On the other hand, according to a health situation analysis carried out and published by the Ministry of Health and Social Protection (MINSALUD) in 2018 [2], around 25.7% of the adult population in this region suffers from hypertension problems. According to the World Health Organization (WHO), noncommunicable diseases such as cancer, diabetes and hypertension are the main death causes worldwide [3]. Among them, hypertension has become the leading cause of death not only in Colombia but around the world, where it has reached up to 1.13 million sick people. In this context, the development of functional foods to prevent cardiovascular risk from highly consumed local fruits is a challenge.

Lulo (*Solanum quitoense* Lam.) is one of the most important tropical fruits in Colombia. Known as naranjilla in Ecuador and as lulum in Peru, this crop from the Solanaceae family is grown in Colombia, Ecuador, Peru, Venezuela, Guatemala, Mexico, Costa Rica, Dominican Republic, and Panama. The plant produces a spherical fruit with a diameter that varies between 3 and 8 cm [4]. In the last years, the lulo fruit has raised considerable interest in the global market due to its organoleptic characteristics, pleasant aroma, and acidic and refreshing taste [5]. Consequently, its exportation has undergone a remarkable increase, particularly reaching the United States and Europe. The fruit has considerable nutritional potential due to its high content of vitamins such as thiamin, riboflavin, vitamin A, proteins, minerals, and spermidine [6]. A recent study has demonstrated the potential antihypertensive use of lulo [7]. For the first time, these authors found N^1,N^4,N^8-tris (Dihydrocafeoil) spermidine and N^1,N^8-bis-(Dihydrocafeoil) spermidine to be bitter bioactive amines in lulo fruit samples. These compounds confer some functional characteristics to lulo for the control of hypertension.

Technological transformation processes are needed to produce innovative fruit derivatives with a long shelf life and, as much as possible, preserved functional attributes and acceptable physical and sensorial characteristics. In this sense, some nonthermal technologies allow for the improvement or maintenance of relevant properties. Compared to traditional heat treatments, the application of high- and/or moderate- pressure homogenization (HPH) to fruit juices has been shown to be less destructive for food compounds of low molecular weight. Likewise, this result may be related to sensory and nutritional qualities and to the sufficient inactivation of different microorganisms. Moreover, it can improve the efficiency and performance of other processes when applied as pretreatment [8,9].

Additionally, food engineering operations like vacuum impregnation (VI) allow for the incorporating of physiologically active compounds to fruit matrices; the structure of which performs as a natural protection, thus resulting in fresh functional foods [10,11]. Fruit juices with improved properties have been incorporated into adequate fruit matrices, leading to functional foods with better composition and high potential to soften some health problems [12].

Due to the valuable and characteristic aroma of lulo fruit and its richness in antioxidant components, some studies regarding volatile profile, antioxidant capacity, and composition in carotenoids, polyphenols, or other micronutrients have been done. In addition, it is possible to find some works in the literature focused on pulp-drying to obtain a powder for industrial use. However, as far as the authors know, no work has been carried out determining the possibilities of using the fruit in VI treatments or analyzing the effect of HPH on the physicochemical and antioxidant properties of juice.

Considering everything mentioned above, it was considered relevant to study physicochemical and antioxidant properties (antioxidant capacity, total phenols, flavonoids and spermidine contents, and polyphenolic compounds profile) of the lulo fruit and its juice. Additionally, VI properties of the fruit and the effect of HPH (50, 100, and 150 MPa and one pass) on the lulo juice properties were studied too. The end goal is to provide knowledge contributing to the use of the indigenous agrifood resources of the region of Chocó (Colombia) in order to improve the health of its population.

2. Materials and Methods

2.1. Food Materials and Sample Preparation

Colombian fresh lulo fruits (*Solanum quitoense* Lam.) were purchased at the Central Market of Valencia (Spain) and stored at 4 °C until processing.

The whole fruits were processed as follows for their analysis: after removing the peduncle, they were washed and cut into pieces, which were crushed in a mortar. This whole crushed fruit is hereafter called "fruit".

The procedure for obtaining the lulo juice was as follows: after removing the peduncle, the whole fresh fruit was washed and blended (Phillips Advance Collection Standmixer, 800 W, 2 L) for 10 min. Then, the blend was filtered and sieved with a 500 µm stainless-steel mesh sift (200/50, CISA). The filtered liquid thus obtained is hereafter called "juice".

The sample preparation for VI experiments was as follows: the fruit was washed and cut transversally into 5 mm thickness and 64 mm diameter slices (FAGOR CF-150 slicer). The fruit skin was not removed, so as to prevent any absorption of materials through the lateral surface of the slices, which allowed assuming unidirectional matter flow. The impregnation experiments were carried out on three lulo fruit batches at a very similar ripeness stage, and three samples per batch were characterized.

Figure 1 summarizes in a flow chart the sample preparation and treatments.

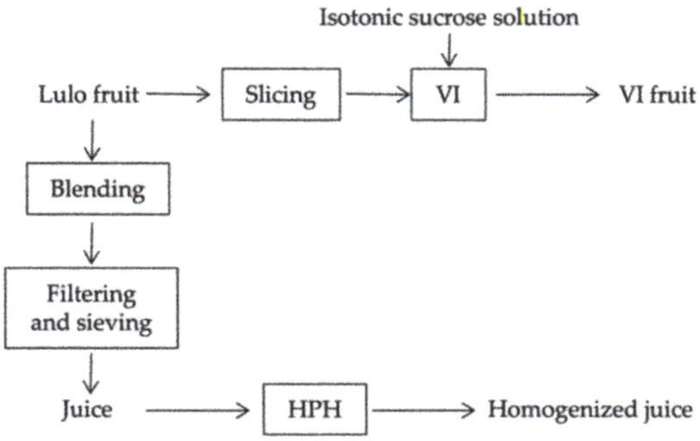

Figure 1. Flow chart summarizing sample preparation and treatments.

2.2. HPH Treatment

Juice samples were homogenized with one pass (Gea Niro Soavi-Panda Plus 2000 homogenizer, Parma, Italy) at 50, 100, and 150 MPa. Homogenized and nonhomogenized juices were refrigerated at 4 °C under aseptic conditions until later analysis.

2.3. VI Experiments

The VI experiments were carried at a pilot plant scale in the Institute of Food Engineering for Development at Universitat Politècnica de València (Spain). The equipment used was described by Fito et al. [13]. In all trials, the sliced fruit samples were immersed in isotonic sucrose aqueous solutions (a_w = 0.994 ± 0.003), and a vacuum pressure of 50 mbar was applied for 10 min. Subsequently, atmospheric pressure was restored, keeping the fruit slices immersed in the impregnation liquid for 10 min more. The impregnation parameters allow us to quantify the volumetric fraction of liquid incorporated into the porous structure and the volumetric deformation of the sample after the vacuum stage (X_1 and γ_1) and at the end of the process (X and γ). The effective porosity (ε_e) provides information on the volumetric fraction of pores that are filled during the VI experiments. All the parameters were calculated using the model equations proposed by Fito et al. [14].

2.4. Physicochemical Characterization

In fruit, the moisture content was quantified by vacuum drying at 60 °C until constant weight [15]. Water activity was measured using a dew point hydrometer (DECAGÓN Aqualab CX-2, ±0.003, Pullman, WA, USA). Brix were determined in a refractometer (ABBE ATAGO BT, NAR T3, Tokyo, Japan) at 20 °C. pH values were measured with a potentiometer (Mettler Toledo Inlab, Schwerzenbach, Switzerland) at 20 °C. The apparent density was obtained through the volume displacement method, using a solids pycnometer and toluene as reference liquid.

In homogenized and nonhomogenized juice, density was determined with a liquid's pycnometer. Particle size distribution was determined using a Malvern Mastersize 2000 system (Malvern Instruments Limited, Worcesterhire, UK) equipped with a blue light source (470 nm wavelength; 0.02–200 micron measuring range). A small amount of sample was diluted in deionized water in the diffractometer cell under moderate agitation until it reached 8–9% darkness. The refractive index values of the juice (cloud) and the dispersant (water) were 1.5 and 1.33, respectively. These measurements were taken using a short-wavelength blue light source in conjunction with forward and backscatter detection to enhance sizing performance in the 0.01–1000 μm range. The particle size distribution of the juice was characterized by percentages in volume (D[4,3]) and in area (D[3,2]) based diameters, and by percentiles d_{10}, d_{50}, and d_{90}, which represent the characteristic diameters under which 10%, 50%, and 90% of the particles are within the distribution. Each analysis was repeated ten times. The rheological behavior was determined by obtaining a flow curve from a rotary rheometer (HAKKE RheoStress 1—RS1 Thermo Electron Corporation, Karlsruhe, Germany), using a Z34 DIN coaxial cylinder sensor system and a temperature bath at 20 ± 1 °C (HAKKE Phoenix 2 controller, Thermo Electron Corporation, Kalsruhe, Germany). The samples were subjected to three ascending and three descending sweeps with a velocity gradient from 0 to 300 s^{-1}. Since all the samples exhibited a non-Newtonian pseudoplastic behavior, samples' flow behaviors were modeled using the Ostwald–de Waele model.

$$\sigma = K \cdot \dot{\gamma}^{n}$$

The parameters K (consistency index ($Pa \cdot s^n$)) and n (flow behavior index (dimensionless)) for the model were obtained by regression using the software HAAKE RheoWin Data Manager v.3.61.0004. The results stated are the average of triplicates.

Color coordinates were obtained through a reflection spectrum between 400 and 700 nm, using a MINOLTA brand spectrocolorimeter (Model CM-3600D, Minolta, Osaka, Japan) with D65 illuminant and a 10 °C observer as references. The resulting CIE-L*a*b* color coordinates allowed for the calculating of the psychometric coordinates: tone (h*ab) and chrome (C*ab). The color difference (ΔE) between each homogenized juice and the non-homogenized one (reference) was calculated. All determinations were made by triplicate.

2.5. Antioxidant Properties

Antioxidant capacity by DPPH and ABTS methods, total phenols, flavonoids and spermidine content, and polyphenolic compounds profile of the lulo fruit and homogenized and nonhomogenized juices were determined.

For determination of total phenols and flavonoids content and antioxidant capacity by DPPH and ABTS method, antioxidants were extracted from fruit or from juice by diluting the samples in an 80:20 (*v/v*) methanol-water solution at a 1:10 ratio (*w/v*) and centrifuged at 10,000 rpm and 20 °C for 5 min (Selecta, "Medifriger BL-S"). Subsequent analyses were carried out on the supernatant (extract) by triplicate.

Total phenol content was determined following the Folin–Ciocalteu method [16,17]. For this procedure, 0.125 mL of extract, 0.125 mL of Folin–Ciocalteu reagent (Sigma-Aldrich, Saint Louis, MO, USA), and 0.5 mL of double-distilled water were mixed and allowed to react for 6 min. After that, 1.25 mL of 7% (*w/v*) sodium carbonate solution and 1 mL of double distilled water were added. Absorbance was measured in a spectrophotometer (Thermo Scientific, Helios Zeta U/Vis, Loughborough, UK) at 765 nm after 90 min. A standard gallic acid curve ranging from 0 to 500 mg/L was obtained. Results were expressed in milligrams of gallic acid equivalent (GAE) per gram of sample.

Flavonoid content was determined following the method described by Luximon-Ramma et al. [18]. In this case, 1.5 mL of extract and 1.5 mL of a 2% (*w/v*) aluminum chloride solution in methanol were mixed and left in the dark for 10 min. Absorbance was measured on a spectrophotometer (Thermo Scientific, Helios Zeta U/Vis, Loughborough, UK) at 368 nm. The resulting data were compared to a standard quercetin curve ranging from 0 to 350 mg/L. The results were expressed in milligrams of quercetin equivalent (EQ) per gram of sample.

Antioxidant capacity by DPPH (2.2-diphenyl-1-picrylhydrazyl) radical was determined following the method described by Kuskoski et al. [19] and Stratil et al. [20], with some modifications. A blend made up of 0.1 mL of the extract, 0.9 mL of methanol, and 2 mL of a 100 μM methanol–DPPH (39.4 μg/mL) solution was prepared. After 60 min of reaction time, absorbance was measured at 517 nm in a spectrophotometer (Thermo Scientific, Helios Zeta UV/Vis, Loughborough, UK). The results were expressed as milligrams of Trolox equivalent (TE) per gram of sample, using the Trolox calibration curve within a 0 to 500 mg/L concentration range.

Antioxidant capacity by ABTS (2.2′-azino-bis-3-ethylbenzothiazoline-6-sulfonic acid) radical was evaluated following the method described by Re et al. [21]. A solution containing 7 mM of ABTS radical and 2.45 mM of potassium persulfate was prepared and left in the dark at room temperature for 16 h. $ABTS^+$ was mixed with phosphate buffer to reach an absorbance of 0.70 ± 0.02 at 734 nm. Then, 0.1 mL of extract was added to 2.9 mL of $ABTS^+$ solution. Absorbance was measured at 734 nm in a spectrophotometer (Thermo Scientific, Helios Zeta UV/Vis, Loughborough, UK) after 0, 3, and 7 min of reaction time. The results were expressed as mg of Trolox equivalent (TE) per gram of sample.

Polyphenolic compounds profile and spermidine content were determined by liquid chromatography—mass spectrometry (LC-MS) analysis. First, phenolic compounds were extracted according to the procedure described by Rodrigues et al. [22] and Svobodova [23], with some modifications. In brief, 5 g of sample were mixed with 20 mL of methanol/water (80:20 *v/v*) solution by stirring (Ultra-Turrax, Staufen, Germany) at 150 rpm and room temperature for 1 h. This mix was centrifuged (Beckman Coulter AvantiTM J-25, Hamburg, Germany) at 3864× *g* and 20 °C for 5 min and the supernatant taken. The extraction procedure was repeated five times. Finally, the supernatant was filtered using a Whatman No. 1 paper filter and, subsequently, a 0.45 µm nylon filter and then directly injected into the HPLC equipment.

The equipment used for separation and identification of phenolic compounds was an (Agilent 1290 HPLC Technologies series infinity System LC, Santa Clara, CA, USA) system with a MS detector and a C18 (1.7 µm, 2.1 × 50 mm, Waters) UHPLC (Ultra High-Performance Liquid Chromatography) column. A flow rate of 0.4 mL/min and an injection volume of 5 µL at 30 °C were applied. The solvents employed were 0.1% formic acid in water (A) and 0.1% formic acid in methanol (B). The applied gradient elution was 10% B (0 min), 10% B (5 min), 100% B (12 min); 100% B (18 min), 10% B (18.5 min), 10% B (25 min). An automated calibration was performed using an external calibrant delivery system (CDS) which infuses calibration solution prior to sample introduction. The selected system was an (AB SCIEX Triple TOF™ 5600 MS, Santa Clara, CA, USA) which was used for data acquisition in both positive and negative modes over a mass range of 80–1000 m/z, under the following conditions: both negative and positive ion modes; ion source gas 1 (GC1): 55 psi; ion source gas 2 (GC2): 55 psi; gas curtain 1:25 psi at 400 °C; negative ion spray voltage (ISVF): −4500; collision energy (CE): −50; positive ion spray voltage (ISVF): 5500; collision energy (CE): 30; accumulation time for both positive and negative modes set at 100 ms. The MS used the IDA (Information-Dependent-Acquisition) acquisition method, the survey scan type (TOF-MS, Time of Fligh-Mass- Spectrometry) and the dependent scan type (product ion) with collision energy set at 50 V/30 V. Likewise, spermidine was quantified using the LC-MS/MS (Liquid Chromatography—Mass Spectrometry) method by means of direct extrapolation on the standard curve. The results were expressed in mg/L. Determinations were done by triplicate.

2.6. Statistical Analysis

The statistical analysis of the data was performed in a Statgraphics Centurion XVII software package, making use of simple or multifactorial analysis of variance (ANOVA) at a 95% confidence level ($p < 0.05$).

3. Results and Discussion

3.1. Vacuum Impregnation Properties of Lulo Fruit

Lulo fruit impregnation parameters provide information on the volume of external liquid that can be incorporated into the fruit tissue by controlled VI. This, in turn, informs us on the viability of incorporating protectants, preservatives, physiologically active compounds, or other additives in the porous structure of the fruit, aiming at its preservation or the formulation of new functional foods [10]. According to Fito et al. [14], the physicochemical properties of the impregnation liquid (mainly viscosity) affect the impregnation parameters, but it is the structural features of the impregnated tissue that are decisive. It is necessary that the tissue have sufficiently big intercellular and hollow spaces which, upon slicing, ensure the presence of open pores and, hence, the flow of the impregnating liquid into the porous structure of the fruit.

Table 1 shows the mean and standard deviation values of the VI parameters corresponding to three different lulo batches impregnated with an isotonic sucrose aqueous solution.

Table 1. Vacuum impregnation parameters of lulo fruit slices (mean ± standard deviation).

Batch	a_w	°Brix	X_1	γ_1	X	γ	εe
1	0.995 ± 0.001 [a]	8.73 ± 0.06 [a]	5 ± 7 [a]	5 ± 4 [a]	8.8 ± 1.6 [a]	3 ± 3 [a]	6 ± 4 [a]
2	0.996 ± 0.003 [a]	8.93 ± 0.06 [a]	2 ± 4 [a]	5 ± 2 [a]	11 ± 2 [a]	3.7 ± 0.9 [a]	8 ± 2 [a]
3	0.994 ± 0.002 [a]	8.70 ± 0.17 [a]	2.5 ± 1.3 [a]	7.1 ± 1.0 [a]	8.6 ± 0.9 [a]	2.9 ± 0.8 [a]	6.3 ± 1.2 [a]

Different letters in the same column indicate statically significant differences ($p \leq 0.05$).

The obtained values reveal the technical feasibility of this type of unitary operation on lulo fruit. No significant differences ($p > 0.05$) were observed in parameters X_1, γ_1, γ and ε_e, while they were observed indeed in parameter X. A linear relation ($R^2 = 0.736$) between Brix and X can be observed, allowing one to state that the riper the fruit, the higher the impregnation of its porous matrix.

The observed differences are fundamentally due to the morphological and structural variability of the fruit, which certainly deserves attention when it comes to the VI process. Positive average values for parameters X_1 (1–5%) and X (8.6–16%) in all batches indicate the incorporation of the impregnation liquid into the porous structure of the fruit during the vacuum stage and total process, respectively. Likewise, the positive volume deformation records registered during the vacuum stage and total process, respectively, expressed by γ_1 (3.9–7.1%) and γ (2.9–6.6%), indicate a volumetric expansion of the fruit matrix, mainly affected by vacuum application [24].

The effective porosity (ε_e) provides information on the volumetric fraction of pores that are filled during the VI experiments. It shows favorable values between 6–9%, which makes the lulo matrix appropriate for the VI process. Interesting observations result from comparing these results to the values obtained for several fruits and vegetables. The current lulo results are much lower than those reported by Betoret et al. [9] and Fito et al. [25] for Granny Smith apple (21 ± 0.9) and Soraya eggplant (64.1 ± 2), but higher than those of Chandler strawberry (6.4 ± 0.3), Hayward kiwi (0.7 ± 0.5), and Bulida apricot (2.2 ± 0.2). Regarding a likely significant batch effect (a_w and Brix) on ε_e, it can be observed that there are no significant differences due to the apparent homogeneity between the studied fruit samples. Differences in Brix and a_w may be associated with different degree of ripeness that bring with them a different structural behavior of the samples during the impregnation process. In this way, selecting fruits with the same degree of ripeness would allow obtaining fruits with a very homogeneous response to the impregnation process.

3.2. Physicochemical Characterization

Table 2 shows the water content, Brix, water activity, pH and density values of the studied fruit, and homogenized and nonhomogenized juice samples.

The only significant differences ($p \leq 0.05$) are observed between fruit and juice soluble solids content (Brix) and density, regardless of HPH treatment. This is due to the fact that some fruit soluble solids remain in the bagasse after juice preparation.

Comparing the density values of homogenized and nonhomogenized juices, a slight increase is produced by the homogenization pressure. This effect can be associated to a decrease in the particle size of the suspended solids and an increase in the stability of the cloud observed in homogenized fruit juices [26–28].

Particle size distribution of homogenized and nonhomogenized juice samples is presented in Figure 2. A monomodal distribution ranging from 1 to 1000 µm can be observed in all juice samples, with a small irregular peak between 10 and 50 µm in the nonhomogenized juice. Although there is a remarkable reduction in particle size as a consequence of homogenization, all curves exhibit a similar particle size distribution pattern.

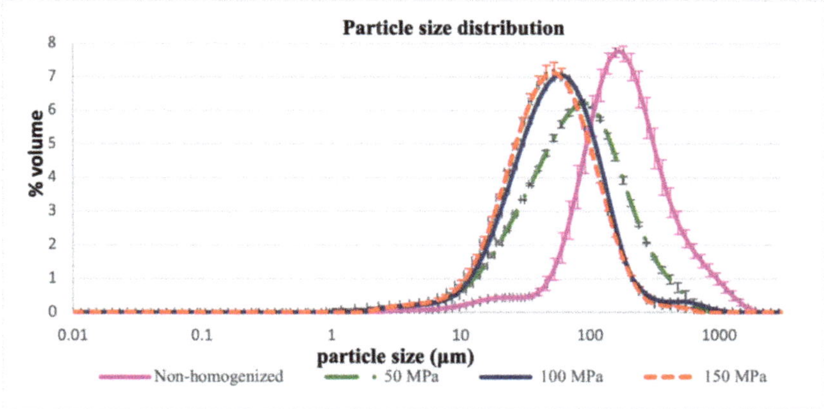

Figure 2. Effect of homogenization pressure on the particle size distribution of homogenized and nonhomogenized juice.

The above is corroborated by statistical analysis of the volume (D[4,3]) and area (D[3,2]) based diameters, which reveals a significantly negative correlation between homogenization pressure and particle size. The maximum and minimum diameters were, respectively, observed in the nonhomogenized juice (251 ± 5 µm) and the one homogenized at 150 MPa (57.94 ± 0.14 µm).

Comparing these results with those reported by Salustiana orange juice (150.1 ± 4.8, 107.7 ± 4.1 µm) and Ortanique mandarin (372.1 ± 1.9, 122.9 ± 2.2 µm) exhibits slightly similar values to those obtained for diameter (D[4,3]) in the present study. However, the (D[4,3]) and (D[3,2]) diameter values found by Castagnini et al. [29] in cranberry juice are much lower. Both authors claim particle size reduction in juices treated with homogenization pressure.

Particle size values below 10%, 50%, and 90% of the particles present in the lulo juices studied in the present work are much higher than those reported by Castagnini et al. [29] for homogenized cranberry juice.

Values of rheological properties show a pseudoplastic behavior in homogenized and nonhomogenized lulo juices (Table 2). Non-Newtonian behavior of fruit juices results from complex interactions between soluble sugars, colloidal pectic substances, and suspended solids. Pseudoplastic behavior reflects a structural reorganization of fluid particles as the velocity gradient increases, not reaching an asymptotic viscosity value. In general, in fruit juices, the higher the soluble solids content, the higher the consistency index. Nonhomogenized lulo juice have a low consistency index (K) and a lower than 1 value of the flow behavior index (n), reflecting the deviation from Newtonian behavior (n = 1) and a good capacity to be pumped and circulated in industrial plants. A significant influence of homogenization pressure on the rheology of the lulo juice as compared to the nonhomogenized one is observed. Yet, no significant differences can be observed across the homogenized samples (50 and 150 MPa). It can also be noted that there is no clear increasing or decreasing trend in the consistency index (K) after increasing the homogenization pressure.

Table 2. Physicochemical characterization, water activity (a_w), moisture content (x_w), (g water/100 g), soluble solids, particle size, rheological properties, and CIE L*a*b* coordinates of the lulo fruit and homogenized and nonhomogenized juice. Mean ± standard deviation of three repetitions. (Different letters in superscripts mean significant differences ($p < 0.05$)).

	Fruit	Nonhomogenized	50 MPa	100 MPa	150 MPa
X_W (%)	91.2 ± 0.4	-	-	-	-
a_w	0.994 ± 0.003 ^{a}	0.994 ± 0.003 ^{a}	0.997 ± 0.003 ^{a}	0.995 ± 0.001 ^{a}	0.996± 0.000 ^{a}
Brix	8.88 ± 0.17 ^{b}	6.57 ± 0.12 ^{a}	6.4 ± 0.4 ^{a}	6.33 ± 0.15 ^{a}	6.4 ± 0.4 ^{a}
pH	3.13 ± 0.16 ^{a}	3.31 ± 0.01 ^{a}	3.12 ± 0.02 ^{a}	3.18 ± 0.03 ^{a}	3.18 ± 0.03 ^{a}
ϱ (g/cm^3)	1.16 ± 0.07 ^{b}	1.036 ± 0.018 ^{a}	1.06 ± 0.04 ^{a,b}	1.07 ± 0.02 ^{a,b}	1.090 ± 0.013 ^{a,b}
			Particle size		
D [4,3]	-	251 ± 5 ^{d}	124 ± 3 ^{c}	75.5 ± 1.2 ^{b}	57.94 ± 0.14 ^{a}
D [3,2]	-	102.3 ± 0.5 ^{d}	49.9 ± 1.5 ^{c}	35.28 ± 0.07 ^{b}	26.83 ± 0.19 ^{a}
d_{10} (μm)	-	75.8 ± 0.3 ^{d}	25.7 ± 1.1 ^{c}	18.94 ± 0.11 ^{b}	15.63 ± 0.03 ^{a}
d_{50} (μm)	-	184.2 ± 1.6 ^{d}	99.82 ± 0.09 ^{c}	60.01 ± 0.11 ^{b}	45.02 ± 0.19 ^{a}
d_{90} (μm)	-	524 ± 17 ^{d}	247 ± 3 ^{c}	153.7 ± 2.0 ^{b}	114.2 ± 0.5 ^{a}
			Rheological properties		
K (Pa.s^n)	-	0.39 ± 0.12 ^{a}	0.9 ± 0.4 ^{b}	0.79 ± 0.02 ^{a,b}	1.3 ± 0.5 ^{b}
n	-	0.44 ± 0.06 ^{b}	0.37 ± 0.04 ^{a}	0.37 ± 0.00 ^{a}	0.34 ± 0.04 ^{a}
R^2		0.99	0.98	0.96	0.98
			Color		
L*	-	40.4 ± 0.4 ^{b}	40.16 ± 0.11 ^{a,b}	39.409 ± 0.012 ^{a}	39.44 ± 0.08 ^{a}
a*	-	9.5 ± 0.2 ^{b}	9.08 ± 0.07 ^{a}	9.15 ± 0.01 ^{a,b}	8.829 ± 0.011 ^{a}
b*	-	34.9 ± 1.4 ^{a}	35.4 ± 0.5 ^{a}	34.54 ± 0.01 ^{a}	33.8 ± 0.2 ^{a}
Cab*	-	36.2 ± 1.3 ^{a}	36.5 ± 0.5 ^{a}	35.73 ± 0.01 ^{a}	34.9 ± 0.2 ^{a}
hab*	-	74.7 ± 0.2 ^{a}	75.61 ± 0.09 ^{b}	75.16 ± 0.01 ^{a,b}	75.37 ± 0.13 ^{a,b}
ΔE	-	-	0.8 ± 0.2 ^{a}	1.250 ± 0.014 ^{a,b}	1.74 ± 0.13 ^{b}

Although sifting the juice drags a considerable amount of soluble solids, when the juice is subjected to HPH, particle size is reduced, the stability of suspended solids increased, and the juice behavior modified as if it increased in soluble solids content (higher K value). While some authors have reported similar results in terms of particle size reduction due to homogenization, others have found contrasting results for factor K. In studying the effect of homogenization on the properties of mixed peach-and-carrot juice, the authors of [26] observed a drop in the consistency index (K) and the flow behavior index. Similar trends were observed by Leite et al. [30] in orange juice and by Silva et al. [31], who studied the effect of homogenization on pineapple pulp, finding that it reduced pseudoplastic behavior (i.e., it increased the flow behavior index n and reduced the index of consistency K). Probably accounting for these contrasts, the authors of [32] have shown that the cell walls of each plant behave differently when subjected to HPH. That is to say, each fruit juice requires a different shear effort, suggesting that HPH may produce contrasting effects on different products, being mainly conditioned by the chemical nature of the components that are suspended in the juice. In studying blueberry juice, Castagnini et al. [29] obtained a consistency index of 0.57 ± 0.03 Pa.s^n and a flow behavior index of 0.33 ± 0.02. Chiralt et al. [33] reported a K value of 2 Pa.s^n and an n value of 0.43 for tomato juice with 12.8% solids and a K value of 6.48 Pa.s^n and an n value of 0.74 for concentrated orange juice homogenized at high pressures. According to these reported data, the consistency of homogenized and nonhomogenized lulo juice is similar to that of blueberry juice, which is of commercial use.

Table 2 shows the parameters L*, a*, b*: L* for perceptual lightness, and a* and b* for the four unique colors of human vision. Coordinate a* is relative to the green–red opponent colors, with negative values toward green and positive values toward red. The b* coordinate represents the blue–yellow opponents, with negative numbers toward blue and positive toward yellow. In addition, the psychrometric coordinates chrome (Cab*), hue (hab*) and ΔE (total color change), as functions of homogenization pressure ranging from homogenized to nonhomogenized lulo juice have been included. A slight reduction can be observed in all parameters under increased homogenization pressure.

The analysis of variance revealed a significant effect of homogenization pressure on all variables except for b* and Cab*, and L* values show significant differences between the nonhomogenized juice and the homogenized ones, the former being slightly brighter than the latter. As for parameter a*, the nonhomogenized juice exhibits the highest value. On the other hand, hue (hab*) exhibits differences between the nonhomogenized juice and the one homogenized at 50 MPa, the former one showing a less orange coloration. No significant differences were observed between the values of the studied color parameters among the homogenized juices. Thus, it can be said that juice color was not affected by pressure intensity.

Global color differences were found between the juice homogenized at 150 MPa and the nonhomogenized one used as reference. However, the visual perception of the color changes in the analyzed juices, described by the ΔE value, was not appreciated.

3.3. Antioxidant Properties

Figure 3 show the effect of the homogenization pressure on total phenol content, flavonoid content, and antioxidant capacity by DPPH• and ABTS radical methods.

The DPPH• method is more sensitive to hydrophobic flavanones, while the $ABTS^+$ method is more sensitive to hydrophilic antiradicals [34]. The $ABTS^+$ free radical decoloration and the DPPH• free radical-scavenging assays have been reported as useful tools to evaluate antiradical activities of different fruits [9].

In comparing obtained values through statistical analysis, significant total phenol differences can be observed between the nonhomogenized juice and that homogenized at 150 MPa. Juices homogenized at 50 MPa and 100 MPa show intermediate values approaching the value of the fruit with increasing the pressure. Total flavonoids content present significant differences between fruit and juices, but this difference is more pronounced when applying the homogenization treatment.

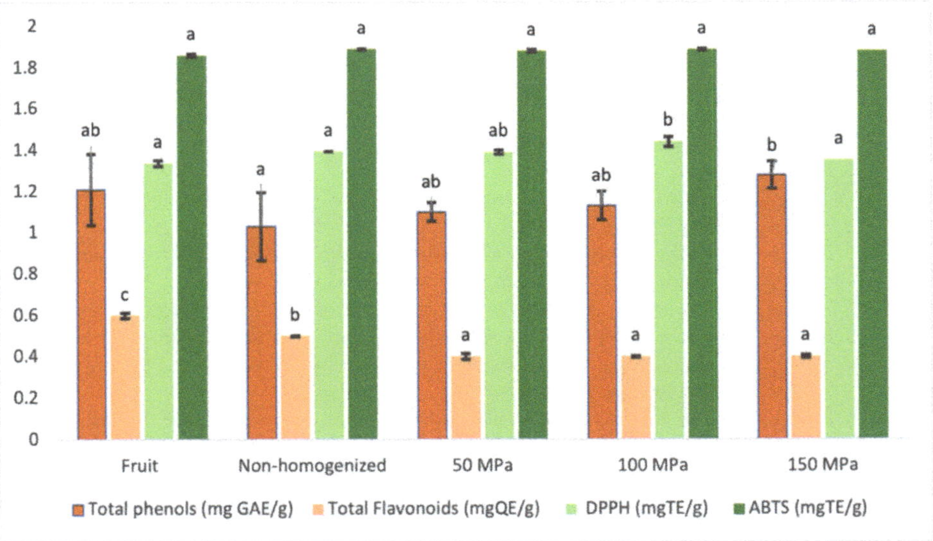

Figure 3. Total phenolic content, total flavonoids content, and antioxidant capacity of lulo fruit and nonhomogenized and homogenized lulo juices by the DPPH (2.2-diphenyl-1-picrylhydrazyl) and $ABTS^+$ (2.2′-azino-bis-3-ethylbenzothiazoline-6-sulfonic acid) methods. Different letters for the same determination indicate significant differences ($p \leq 0.05$).

In both cases, obtaining juice decreases the content of total phenols and flavonoids from fruit. It would be associated with the removal of the solid fraction, especially from the skin which is very rich in these compounds [35]. However, the application of an HPH produces a phenol recovery, as the pressure value applied increases while the same treatment reduces flavonoids content. As it has been explained by other authors [36], food matrix properties and processing can promote a different effect on bioactive compounds. The forces and temperature stresses created in the homogenization valve in HPH treatment could lead to a degradation of flavonoid during homogenization in this case. However, in those cases in which degradation of bioactive compounds has not been occurred yet, particle size decrease can facilitate the extraction, as it is observed in the case of phenolic compounds.

The values found in the current fruit samples do not differ much from those obtained by Contreras-Calderón et al. [1] (0.583 ± 0.024 mg GAE/g, fresh sample); by Igual et al. [37] (0.811 ± 0.016 mg GAE/g, fresh sample); and by Vasco et al. [38] (0.91 ± 0.17 mg GAE/g, fresh sample), also in lulo. On the other hand, the flavonoid content values reported by Igual et al. [37] were 0.16 ± 0.02 mg RE equivalents/g, which is a lower value than those obtained in this study (mg of quercetin equivalent).

The results of antioxidant capacity vary from 1.44 ± 0.02 mg TE/g (82.40 ± 1.41% inhibition) to 1.35 ± 0.06 mg TE/g (76.8 ± 0.8% inhibition) for the assays by DPPH$^{\bullet}$; and from 1.88 ± 0.03 mg TE/g (89.0 ± 1.5% inhibition) to 1.86 ± 0.01 mg TE/g (87.9 ± 0.3% inhibition) for the ABTS^{+} assays.

The values obtained by the ABTS^{+} method in both fruit and juice samples do not show significant differences. However, by the DPPH$^{\bullet}$ method, the juice homogenized at 100 MPa shows the highest value being significantly different from all other samples. The decrease observed in the content of total phenol and flavonoid contents in the juice compared with the fruit is not reflected when analyzing the antiradical activity. This could be explained by the fact that the liquefaction and HPH treatment may cause unbounding or chemical changes in some phenolic compounds of the fruit, giving rise to other compounds with antiradical capacity that compensate for the loss associated with sieving [39].

Authors like Forero et al. [5] and Contreras-Calderón et al. [1] in analyzing the same fruit have reported values of 71.0 ± 2.3 mg TE/g solid and 3.05 ± 0.21 mg TE/g fresh weight, respectively. The former authors obtained a fairly high value, while the latter reported a value even higher value than the one found in this study for the fruit and homogenized and nonhomogenized juices. Contrastingly, in applying the DPPH$^{\bullet}$ method, Vasco et al. [38] found records of 0.80 ± 0.22 mg TE/g of fresh sample, with this value being lower than those reported in this study, which vary from 1.34 to 1.44 mg TE/g of sample.

3.4. Profile of Phenolic Compounds by High-Performance Liquid Chromatography Coupled to Mass Spectrometry (LC-MS/MS)

Figure 4 shows the percentage distribution class of polyphenolic compounds in the lulo fruit and its homogenized and nonhomogenized juices.

A greater variety of phenolic components has been identified in homogenized juices than in nonhomogenized ones, and this variety increases with increasing homogenization pressure. Results are consistent with the results on the total phenol content previously shown in Figure 3.

In general, hydroxycinnamic acids and flavonoids were the main phenolic subclass found. It is coherent with a previous study reported by Gancel et al. [6] showing that hydroxycinnamic acids and flavonoids (i.e., quercetin glycosides, kaempferol derivatives, and 5-*O*-caffeoylquinic acid) are the dominant phenolic compounds in all parts of the lulo fruit [6].

Suárez-Jacobo et al. [40] also detected an increase in hydroxycinnamic acids in clarified apple juice when pressures of 100, 200, and 300 MPa were applied, although the general differences between the studied antioxidant activity assessment methods were not significant. Velázquez-Estrada et al. [41] have shown that orange juice homogenized at 200 and

300 MPa increased the content of flavonoids which, in this beverage with high hesperidine content, precipitate forming crystals that intertwine with proteins and other compounds.

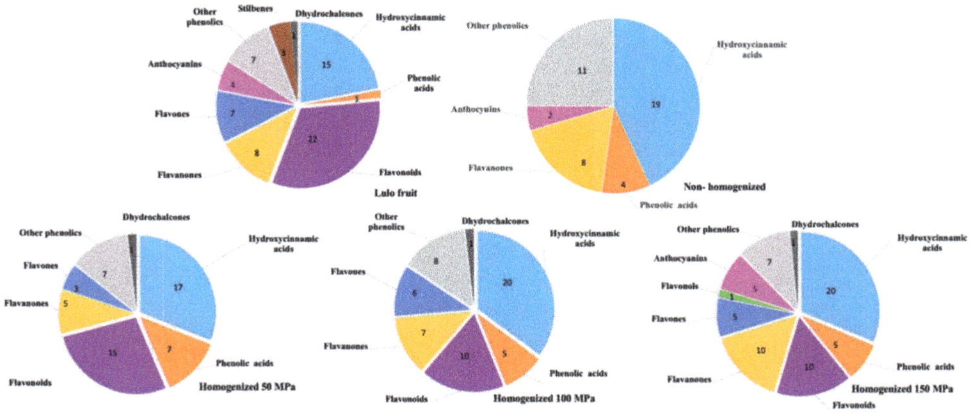

Figure 4. Percentage distribution of phenolic compounds identified in lulo fruit, nonhomogenized, and homogenized lulo juice.

On the other hand, it is important to mention the absence of flavonoids in the nonhomogenized lulo juice. Nevertheless, flavanones have been identified. It can be attributed to the fact that, during its elaboration, some phenolic groups (flavonoids) are bound to sugars or other molecules, thus remaining integrated in the cloud fraction and making them less accessible. Therefore, their release requires physical phenomena related to, in this case, the HPH treatment [8,42–44].

Some studies have shown that the Solanaceae family, especially the genus *Solanum*, are rich sources of antioxidants, such as phenolic compounds and flavonoids, and alkalis. The presence of these compounds in food has a positive impact, mainly as free radical scavengers. In addition to their nutritional benefits, they have also been associated with various biological activities such as anti-inflammatory, antitumor, and benefits related to cardiovascular diseases that include hypertension [45–48].

Tables S1–S6, included as supplementary files present the identified compounds by LC-MS/MS in the fresh fruit extracts and their nonhomogenized and homogenized juices at different pressures and their retention time, experimental m/z, theoretical mass, molecular formula, and MS/MS fragment data. A SCIEX OS software was used for the neutral molecule and error (ppm) data, which were compared to the literature. Regarding LC-MS/MS, a total of 288 compounds were identified, including 91 hydroxycinnamic acids, 22 phenolic acids, 57 flavonoids, 38 flavanones, 21 flavones, 1 flavanol, 40 anthocyanins, 7 other phenolics, 3 stilbenes, and 4 dihydrochalcones. Qualitatively speaking, flavonoids (flavanones, flavons, and flavonols) represent the main phenolic class in this analysis, wherein 116 compounds were found in the fresh fruit and its juices homogenized at different pressures (50, 100, and 150 MPa). For their part, hydroxycinnamic acids were the second class, followed by other phenolics and phenolic acids. Hence, it can be said that fresh lulo fruit contains phenolic compounds of interest.

Table S6 shows $[M-H]^-$-derived ions, MS^2 fragments, and molecular formula for identified compounds. Compound 1 was identified as caffeic acid with $[M-H]^-$ 179 and MS^2 fragment m/z 135, 134, 89, reported by other authors [49–51]. Compounds 2, 3, and 4 presented the same molecular ion $[M-H]^-$ 353 with a MS^2 fragment m/z 191/179. These were described by [6,23,52–56].

Previous studies Park [57] showed that the presence of 5-caffeoylquinic acid and caffeic acid in fruits reduces the risk of suffering cardiovascular diseases through the suppression of *p*-selectin.

Compounds 5, 6, and 7 were assigned as *p*-coumaroylquinic acid, 4-*p*-coumaroylquinic acid, and 5-*p*-coumaroylquinic acid, respectively, with molecular mass $[M-H]^-$ = 337 and with a MS^2 fragment at m/z 191,163 as reported by Pereira et al. [58], Gutiérrez Ortiz et al. [59], Brahem et al. [52], Mikulic-Petkovsek et al. [60], Gómez-Romero et al. [55], and Rodríguez-Medina et al. [61]. Compounds 8 and 9 were, respectively, identified as quercetin-3-O-rhamnoside ($[M-H]^-$ = 447 MS^2 fragment at m/z 301) and quercetin-3-*O*-rutinoside ($[M-H]^-$ = 606 MS^2 fragment at m/z 300). These compounds have also been reported by Liu et al. [62], Kolnaik-Ostek et al. [53], Dorta et al. [63], and Fu et al. [64]. Compounds 10 and 11 correspond to isorhamnetin-3-O-rutinoside; isorhamnetin-3-*O*-rutinoside, respectively. Likewise, compounds 12 and 13 corresponding to kaempferol 3-*O*-rutinoside and kaempferol 3-*O*-glucoside; compounds 14, 15, and 16 were identified as narirutin, phloridzin, and p-coumaroyl glucose, in that order. Similarly, the compounds coumarin, sesamol, naringenin, malvidin 3-*O*-(6-acetyl-glucoside) were also identified and reported by Brahem et al. [52], Diaz-García et al. [51], Kolniak-Ostek et al. [53], and Phenol-Explorer Database [65]. Baret et al. [66] reported that polyphenols and flavonoids, such as resveratrol, quercetin, epigallocathechin-3-gallate, and curcumin, help reduce fat storage, blood pressure, blood glucose, and hemoglobin-A1c, as well as reducing insulin resistance. They showed that some compounds such as caffeic acid, gallic acid, myricetin, and catechin protect oxidative stress and help prevent cardiovascular diseases. Furthermore, Asgary et al. [67] reported that consuming pomegranate juice, which has a very similar compound to the lulo fruit juice samples studied, for 2 weeks exerts a positive effect on hypertensive individuals, and these same authors found that pomegranate juice can be considered as an effective complement to antihypertensive medications and as a component of the daily regimen for patients at high risk of hypertension.

Table S6, also included as a supplementary file, show the polyamines identified in fruit and juice samples. The same compounds were identified by Svobodova et al. [23] in peaeggplant, Wu et al. [68] in eggplant, and Rodrigues et al. [22] in mana-cubiu.

Forero et al. [7] demonstrated the potential of the lulo as an antihypertensive due to its free bioactive amines: N^1,N^4,N^8-tris(dihydrocaffeoyl) spermidine and N^1,N^8-bis(dihydrocaffeoyl) spermidine. The inhibitory activity of the fresh and dried fruit with angiotensin I-converting enzyme confirmed beneficial effect on hypertension. In another study, Gancel et al. [6] identified the same bioactive compounds in the lulo.

Polyamines are present in food, such as milk and some plants, taking part in a wide range of biologically process, such as, cellular proliferation, free radicals scavenging, the differentiation of immune cells, and neurotransmission [69].

Bomtempo et al. [70] evaluated the presence of polyamines and other bioactive amines in four varieties of the passion fruit species, reporting high spermidine concentrations and emphasizing the potential of the passifloras with functional properties relevant for the plant and human health.

However, the analytical determination of bioactive compounds in any food is required but not sufficient. Nutritional and healthy effect of food is determined by its content in macro- and micro-nutrients, their release at the target site in the adequate form, and its suitable assimilation. These three aspects considered together define the functionality of a food and are reflected separately in digestibility, bioaccessibility, and bioavailability properties. It would be necessary to carry out in vitro digestion studies or in vivo studies to quantify these properties which may be affected both by the food matrix and by the treatments applied.

4. Conclusions

The results obtained have shown lulo as a fruit with advantageous physicochemical and functional properties for the development of healthy food products from fresh native crops of the Colombian Pacific region. Lulo fruit can provide the bases of a new fruit juice flavor as well as other products derived from its richness in polyphenols, polyamines, and other antioxidant components.

Its structural characteristics determined indirectly from the impregnation parameters would allow the incorporation of protectants, preservatives, physiologically active compounds, or other additives. This incorporation would be slightly influenced by the maturity index but not by the variability among fruits.

From the phytochemical profile as obtained by LC-MS, 288 compounds belonging to different phenolic classes were found in the fruit and its homogenized and nonhomogenized juices (mainly flavonoids and hydroxycinnamic acids). Increasing pressure of HPH treatment increase the diversity in polyphenols from juice. Additionally, bioactive amines such as N^1,N^4,N^8-tris(dihydrocaffeoyl) spermidine and N^1,N^8-bis(dihydrocaffeoyl) spermidine, whose effect against hypertension has been shown in previous studies, have been identified in fruit and juice samples.

In relation to the antiradical capacity provided by antioxidant compounds, it decreases in fresh juice compared to fruit, due to the retention of part of the solids in the filtering operation. However, it is worthwhile mentioning that the HPH at 100–150 MPa is adequate for preserving the antiradical capacity of the fruit increasing the antioxidant value of fresh juice.

However, the potential beneficial effect of the lulo fruit or of any of the analyzed juices should be assessed through in vitro studies that provide information on the bioavailability and bioavailability of the analyzed components.

Supplementary Materials: The following are available online at https://www.mdpi.com/article/10.3390/foods10040817/s1; Supplementary data included in Tables S1–S6.

Author Contributions: Conceptualization, N.B.; methodology, L.I.H.-C. and N.B.; writing—original draft preparation, L.I.H.-C.; writing—review and editing, L.I.H.-C., N.B., C.B., and L.S.; funding acquisition, L.I.H.-C. and N.B. All authors have read and agreed to the published version of the manuscript.

Funding: Authors thank the grant provided to Leidy I. Hinestroza by Technological University of Chocó-Colombia [Fortalecimiento de los Encadenamientos Productivos de las Subregiones del Chocó. BPIN 2013000100284].

Institutional Review Board Statement: Not applicable.

Informed Consent Statement: Not applicable.

Conflicts of Interest: The authors declare no conflict of interest.

References

1. Contreras-Calderón, J.; Calderón-Jaimes, L.; Guerra-Hernández, E.; García-Villanova, B. Antioxidant capacity, phenolic content and vitamin C in pulp, peel and seed from 24 exotic fruits from Colombia. *Food Res. Int.* **2011**, *44*, 2047–2053. [CrossRef]
2. Ministerio de Salud y Protección Social Análisis de Situación de Salud. Colombia. Available online: https://www.minsalud.gov.co/salud/publica/epidemiologia/Paginas/analisis-de-situacion-de-salud-.aspx (accessed on 25 August 2019).
3. WHO Health Topics: Hypertension. Available online: https://www.who.int/news-room/fact-sheets/detail/hypertension (accessed on 2 January 2020).
4. Ramírez, F.; Kallarackal, J.; Davenport, T.L. Lulo (*Solanum quitoense* Lam.) reproductive physiology: A review. *Sci. Hortic.* **2018**, *238*, 163–176. [CrossRef]
5. Forero, D.P.; Orrego, C.E.; Peterson, D.G.; Osorio, C. Chemical and sensory comparison of fresh and dried lulo (*Solanum quitoense* Lam.) fruit aroma. *Food Chem.* **2015**, *169*, 85–91. [CrossRef] [PubMed]
6. Gancel, A.L.; Alter, P.; Dhuique-Mayer, C.; Ruales, J.; Vaillant, F. Identifying carotenoids and phenolic compounds in naranjilla (*Solanum quitoense* Lam. var. Puyo hybrid), an Andean fruit. *J. Agric. Food Chem.* **2008**, *56*, 11890–11899. [CrossRef] [PubMed]
7. Forero, D.P.; Masatani, C.; Fujimoto, Y.; Coy-Barrera, E.; Peterson, D.G.; Osorio, C. Spermidine derivatives in lulo (*Solanum quitoense* Lam.) fruit: Sensory (taste) versus biofunctional (ACE-inhibition) properties. *J. Agric. Food Chem.* **2016**, *64*, 5375–5383. [CrossRef]
8. Mesa, J.; Hinestroza-Córdoba, L.I.; Barrera, C.; Seguí, L.; Betoret, E.; Betoret, N. High homogenization pressures to improve food quality, functionality and sustainability. *Molecules* **2020**, *25*, 3305. [CrossRef]
9. Betoret, E.; Sentandreu, E.; Betoret, N.; Fito, P. Homogenization pressures applied to citrus juice manufacturing. Functional properties and application. *J. Food Eng.* **2012**, *111*, 28–33. [CrossRef]
10. Betoret, E.; Betoret, N.; Vidal, D.; Fito, P. Functional foods development: Trends and technologies. *Trends Food Sci. Technol.* **2011**, *22*, 498–508. [CrossRef]

11. Betoret, E.; Betoret, N.; Rocculi, P.; Dalla Rosa, M. Strategies to improve food functionality: Structure-property relationships on high pressures homogenization, vacuum impregnation and drying technologies. *Trends Food Sci. Technol.* **2015**, *46*, 1–12. [CrossRef]
12. Betoret, E.; Sentandreu, E.; Betoret, N.; Codoñer-Franch, P.; Valls-Bellés, V.; Fito, P. Technological development and functional properties of an apple snack rich in flavonoid from mandarin juice. *Innov. Food Sci. Emerg. Technol.* **2012**, *16*, 298–304. [CrossRef]
13. Fito, P.; Chiralt, A.; Barat, J.; Martinez-Monzo, J. Vacuum Impregnation in Fruit Processing. In *Food Preservation Technology*; Lozano, J., Añon, C., Parada.Aries, E., Barbosa-Cánovas, P., Eds.; Gustavo Trends in Food Engineering, Technomic Publishing Company: Lancaster, PA, USA, 2000; pp. 149–164. ISBN 1-56676-991-4.
14. Fito, P.; Andrés, A.; Chiralt, A.; Pardo, P. Coupling of Hydrodynamic Mechanism and Deformation-Relaxation Phenomena during Vacuum Treatments in Solid Porous Food-Liquid Systems. *J. Food Eng.* **1996**, *27*, 229–240. [CrossRef]
15. AOAC. *Official Method 973.18 Official Methods of Analysis of the Association of Official Analytical Chemists*, 18th ed.; Association of Official Analytical Chemist: Arlington, VA, USA, 2000.
16. Singleton, V.L.; Orthofer, R.; Lamuela-Raventós, R.M. Analysis of total phenols and other oxidation substrates and antioxidants by means of folin-ciocalteu reagent. *Methods Enzymol.* **1999**, *299*, 152–178. [CrossRef]
17. Wolfe, K.L.; Liu, R.H. Apple peels as a value-added food ingredient. *J. Agric. Food Chem.* **2003**, *51*, 1676–1683. [CrossRef] [PubMed]
18. Luximon-Ramma, A.; Bahorun, T.; Soobrattee, M.A.; Aruoma, O.I. Antioxidant activities of phenolic, proanthocyanidin, and flavonoid components in extracts of Cassia fistula. *J. Agric. Food Chem.* **2002**, *50*, 5042–5047. [CrossRef]
19. Kuskoski, E.M.; Asuero, A.G.; Troncoso, A.M.; Mancini-Filho, J.; Fett, R. Aplicación de diversos métodos químicos para determinar actividad antioxidante en pulpa de frutos. *Ciência Tecnol. Aliment.* **2005**, *25*, 726–732. [CrossRef]
20. Stratil, P.; Klejdus, B.; Kubáň, V. Determination of total content of phenolic compounds and their antioxidant activity in vegetables—Evaluation of spectrophotometric methods. *J. Agric. Food Chem.* **2006**, *54*, 607–616. [CrossRef]
21. Re, R.; Pellegrini, N.; Proteggente, A.; Pannala, A.; Yang, M.; Rice-Evans, C. Antioxidant activity applying an improved ABTS radical cation decolorization assay. *Free Radic. Biol. Med.* **1999**, *26*, 1231–1237. [CrossRef]
22. Rodrigues, E.; Mariutti, L.R.B.; Mercadante, A.Z. Carotenoids and phenolic compounds from *Solanum sessiliflorum*, an unexploited amazonian fruit, and their scavenging capacities against reactive oxygen and nitrogen species. *J. Agric. Food Chem.* **2013**, *61*, 3022–3029. [CrossRef]
23. Svobodova, B.; Barros, L.; Sopik, T.; Calhelha, R.C.; Heleno, S.; Alves, M.J.; Walcott, S.; Kuban, V.; Ferreira, I.C.F.R. Non-edible parts of *Solanum stramoniifolium* Jacq.-a new potent source of bioactive extracts rich in phenolic compounds for functional foods. *Food Funct.* **2017**, *8*, 2013–2021. [CrossRef]
24. Cortéz Rodriguez, M.; Herrera Herrera, E.A.; Gil González, J.H. Impregnación de Uchua (Physalis peruviana L.) de forma Semiesférica con una Emunsión Fortificante. *Biotecnol. Sect. Agropecu. Agroind.* **2016**, *14*, 27. [CrossRef]
25. Fito, P.; Chiralt, A.; Barat, J.M.; Andrés, A.; Martínez-Monzó, J.; Martínez-Navarrete, N. Vacuum impregnation for development of new dehydrated products. *J. Food Eng.* **2001**, *49*, 297–302. [CrossRef]
26. Wellala, C.K.D.; Bi, J.; Liu, X.; Liu, J.; Lyu, J.; Zhou, M.; Marszałek, K.; Trych, U. Effect of high pressure homogenization combined with juice ratio on water-soluble pectin characteristics, functional properties and bioactive compounds in mixed juices. *Innov. Food Sci. Emerg. Technol.* **2020**, *60*, 102279. [CrossRef]
27. Cerdán-Calero, M.; Izquierdo, L.; Sentandreu, E. Valencia Late orange juice preserved by pulp reduction and high pressure homogenization: Sensory quality and gas chromatography-mass spectrometry analysis of volatiles. *LWT—Food Sci. Technol.* **2013**, *51*, 476–483. [CrossRef]
28. Betoret, E.; Betoret, N.; Carbonell, J.V.; Fito, P. Effects of pressure homogenization on particle size and the functional properties of citrus juices. *J. Food Eng.* **2009**, *92*, 18–23. [CrossRef]
29. Castagnini, J.M. Estudio del Proceso de Obtención de Zumo de Arándanos y su Utilización Como Ingrediente Para la Obtención de un Alimento Funcional por Impregnación a Vacío. Ph.D. Thesis, Universitat Politécnica de Valéncia, Valencia, Spain, 2014.
30. Leite, T.S.; Augusto, P.E.D.; Cristianini, M. The use of high pressure homogenization (HPH) to reduce consistency of concentrated orange juice (COJ). *Innov. Food Sci. Emerg. Technol.* **2014**, *26*, 124–133. [CrossRef]
31. Silva, V.M.; Sato, A.C.K.; Barbosa, G.; Dacanal, G.; Ciro-Velásquez, H.J.; Cunha, R.L. The effect of homogenisation on the stability of pineapple pulp. *Int. J. Food Sci. Technol.* **2010**, *45*, 2127–2133. [CrossRef]
32. Lopez-Sanchez, P.; Svelander, C.; Bialek, L.; Schumm, S.; Langton, M. Rheology and Microstructure of Carrot and Tomato Emulsions as a Result of High-Pressure Homogenization Conditions. *J. Food Sci.* **2011**, *76*. [CrossRef]
33. Chiralt, A.; Martinez-Navarrete, N.; González, C.; Talens, P.; Moraga, G. *Propiedades Reológicas de Alimentos. Reología de Alimentos Fluidos. En: Propiedades Físicas de los Alimentos*, 1st ed.; Universitat Politécnica de Valéncia-España: Valencia, Spain, 2012; Chapter 4.
34. Del Caro, A.; Piga, A.; Vacca, V.; Agabbio, M. Changes of flavonoids, vitamin C and antioxidant capacity in minimally processed citrus segments and juices during storage. *Food Chem.* **2004**, *84*, 99–105. [CrossRef]
35. Hinestroza-Córdoba, L.I.; Serna, S.D.; Seguí, L.; Barrera, C.; Betoret, N. Characterization of powdered lulo (*Solanum quitoense*) bagasse as a functional food ingredient. *Foods* **2020**, *9*, 723. [CrossRef]
36. Betoret, E.; Calabuig-Jiménez, L.; Patrignani, F.; Lanciotti, R.; Dalla Rosa, M. Effect of high pressure processing and trehalose addition on functional properties of mandarin juice enriched with probiotic microorganisms. *LWT—Food Sci. Technol.* **2017**, *85*, 418–422. [CrossRef]

37. Igual, M.; Ramires, S.; Mosquera, L.H.; Martínez-Navarrete, N. Optimization of spray drying conditions for lulo (*Solanum quitoense* L.) pulp. *Powder Technol.* **2014**, *256*, 233–238. [CrossRef]
38. Vasco, C.; Ruales, J.; Kamal-Eldin, A. Total phenolic compounds and antioxidant capacities of major fruits from Ecuador. *Food Chem.* **2008**, *111*, 816–823. [CrossRef]
39. Nagar, E.; Berenshtein, L.; Okun, Z.; Shpigelman, A. The structure-dependent influence of high pressure processing on polyphenol-cell wall material (CWM) interactions and polyphenol-polyphenol association in model systems: Possible implication to accessibility. *Innov. Food Sci. Emerg. Technol.* **2020**, *66*, 102538. [CrossRef]
40. Suárez-Jacobo, Á.; Rüfer, C.E.; Gervilla, R.; Guamis, B.; Roig-Sagués, A.X.; Saldo, J. Influence of ultra-high pressure homogenisation on antioxidant capacity, polyphenol and vitamin content of clear apple juice. *Food Chem.* **2011**, *127*, 447–454. [CrossRef]
41. Velázquez-Estrada, R.M.; Hernández-Herrero, M.M.; Rüfer, C.E.; Guamis-López, B.; Roig-Sagués, A.X. Influence of ultra high pressure homogenization processing on bioactive compounds and antioxidant activity of orange juice. *Innov. Food Sci. Emerg. Technol.* **2013**, *18*, 89–94. [CrossRef]
42. Minatel, I.O.; Borges, C.V.; Ferreira, M.I.; Gomez, H.A.G.; Chen, C.-Y.O.; Lima, G.P.P. Phenolic Compounds: Functional Properties, Impact of Processing and Bioavailability. In *Phenolic Compounds—Biological Activity*; Soto-Hernandez, M., Palma-Tenango, M., del Rosario Garcia Mateos, M., Eds.; InTech: Nappanee, IN, USA, 2017. [CrossRef]
43. Persic, M.; Mikulic-Petkovsek, M.; Slatnar, A.; Veberic, R. Chemical composition of apple fruit, juice and pomace and the correlation between phenolic content, enzymatic activity and browning. *LWT—Food Sci. Technol.* **2017**, *82*, 23–31. [CrossRef]
44. Capanoglu, E.; De Vos, R.C.H.; Hall, R.D.; Boyacioglu, D.; Beekwilder, J. Changes in polyphenol content during production of grape juice concentrate. *Food Chem.* **2013**, *139*, 521–526. [CrossRef]
45. Morais, M.G.; Saldanha, A.A.; Costa Rodrigues, J.P.; Cotta Mendes, I.; Ferreira, L.M.; Avelar Amado, P.; de Souza Farias, K.; Samúdio Santos Zanuncio, V.; Brentan da Silva, D.; Carmo Horta Pinto, F.; et al. Chemical composition, antioxidant, anti-inflammatory and antinociceptive activities of the ethanol extract of ripe fruits of *Solanum lycocarpum* St. Hil. (Solanaceae). *J. Ethnopharmacol.* **2020**, *262*, 113125. [CrossRef]
46. More, G.K. A review of the ethnopharmacology, phytochemistry and pharmacological relevance of the South African weed *Solanum sisymbriifolium* Lam. (Solanaceae). *Environ. Dev. Sustain.* **2019**, *21*, 37–50. [CrossRef]
47. Rios, R.; da Silva, H.B.F.; Carneiro, N.V.Q.; Pires, A.d.O.; Carneiro, T.C.B.; Costa, R.d.S.; Marques, C.R.; Machado, M.S.S.; Velozo, E.d.S.; Silva, T.M.G.d.; et al. *Solanum paniculatum* L. decreases levels of inflammatory cytokines by reducing NFKB, TBET and GATA3 gene expression in vitro. *J. Ethnopharmacol.* **2017**, *209*, 32–40. [CrossRef] [PubMed]
48. Da Costa, G.A.F.; Morais, M.G.; Saldanha, A.A.; Assis Silva, I.C.; Aleixo, Á.A.; Ferreira, J.M.S.; Soares, A.C.; Duarte-Almeida, J.M.; Lima, L.A.R.D.S. Antioxidant, Antibacterial, Cytotoxic, and Anti-Inflammatory Potential of the Leaves of Solanum lycocarpum A. St. Hil. (Solanaceae). *Evid.-Based Complement. Altern. Med.* **2015**, *2015*. [CrossRef]
49. Ismail, B.B.; Pu, Y.; Guo, M.; Ma, X.; Liu, D. LC-MS/QTOF identification of phytochemicals and the effects of solvents on phenolic constituents and antioxidant activity of baobab (Adansonia digitata) fruit pulp. *Food Chem.* **2019**, *277*, 279–288. [CrossRef]
50. Pinela, J.; Montoya, C.; Carvalho, A.M.; Martins, V.; Rocha, F.; Barata, A.M.; Barros, L.; Ferreira, I.C.F.R. Phenolic composition and antioxidant properties of ex-situ conserved tomato (*Solanum lycopersicum* L.) germplasm. *Food Res. Int.* **2019**, *125*, 108545. [CrossRef]
51. Díaz-García, M.C.; Obón, J.M.; Castellar, M.R.; Collado, J.; Alacid, M. Quantification by UHPLC of total individual polyphenols in fruit juices. *Food Chem.* **2013**, *138*, 938–949. [CrossRef]
52. Brahem, M.; Renard, C.M.G.C.; Eder, S.; Loonis, M.; Ouni, R.; Mars, M.; Le Bourvellec, C. Characterization and quantification of fruit phenolic compounds of European and Tunisian pear cultivars. *Food Res. Int.* **2017**, *95*, 125–133. [CrossRef]
53. Kolniak-Ostek, J. Content of bioactive compounds and antioxidant capacity in skin tissues of pear. *J. Funct. Foods* **2016**, *23*, 40–51. [CrossRef]
54. Chanforan, C.; Loonis, M.; Mora, N.; Caris-Veyrat, C.; Dufour, C. The impact of industrial processing on health-beneficial tomato microconstituents. *Food Chem.* **2012**, *134*, 1786–1795. [CrossRef] [PubMed]
55. Gómez-Romero, M.; Segura-Carretero, A.; Fernández-Gutiérrez, A. Metabolite profiling and quantification of phenolic compounds in methanol extracts of tomato fruit. *Phytochemistry* **2010**, *71*, 1848–1864. [CrossRef] [PubMed]
56. Hokkanen, J.; Mattila, S.; Jaakola, L.; Pirttilä, A.M.; Tolonen, A. Identification of phenolic compounds from lingonberry (Vaccinium vitis-idaea L.), Bilberry (Vaccinium myrtillus L.) andHybrid Bilberry (Vaccinium x intermedium Ruthe L.) Leaves. *J. Agric. Food Chem.* **2009**, *57*, 9437–9447. [CrossRef] [PubMed]
57. Park, J.B. 5-Caffeoylquinic acid and caffeic acid orally administered suppress P-selectin expression on mouse platelets. *J. Nutr. Biochem.* **2009**, *20*, 800–805. [CrossRef] [PubMed]
58. Pereira, A.P.A.; Angolini, C.F.F.; Paulino, B.N.; Lauretti, L.B.C.; Orlando, E.A.; Silva, J.G.S.; Neri-Numa, I.A.; Souza, J.D.R.P.; Pallone, J.A.L.; Eberlin, M.N.; et al. A comprehensive characterization of *Solanum lycocarpum* St. Hill and *Solanum oocarpum* Sendtn: Chemical composition and antioxidant properties. *Food Res. Int.* **2018**, 1–9. [CrossRef]
59. Gutiérrez Ortiz, A.L.; Berti, F.; Navarini, L.; Monteiro, A.; Resmini, M.; Forzato, C. Synthesis of p-coumaroylquinic acids and analysis of their interconversion. *Tetrahedron Asymmetry* **2017**, *28*, 419–427. [CrossRef]
60. Mikulic-Petkovsek, M.; Schmitzer, V.; Jakopic, J.; Cunja, V.; Veberic, R.; Munda, A.; Stampar, F. Phenolic compounds as defence response of pepper fruits to Colletotrichum coccodes. *Physiol. Mol. Plant Pathol.* **2013**, *84*, 138–145. [CrossRef]

61. Rodríguez-Medina, I.C.; Segura-Carretero, A.; Fernández-Gutiérrez, A. Use of high-performance liquid chromatography with diode array detection coupled to electrospray-Qq-time-of-flight mass spectrometry for the direct characterization of the phenolic fraction in organic commercial juices. *J. Chromatogr. A* **2009**, *1216*, 4736–4744. [CrossRef]
62. Liu, P.; Li, L.; Song, L.; Sun, X.; Yan, S.; Huang, W. Characterisation of phenolics in fruit septum of Juglans regia Linn. by ultra performance liquid chromatography coupled with Orbitrap mass spectrometer. *Food Chem.* **2019**, *286*, 669–677. [CrossRef]
63. Dorta, E.; González, M.; Lobo, M.G.; Sánchez-Moreno, C.; de Ancos, B. Screening of phenolic compounds in by-product extracts from mangoes (Mangifera indica L.) by HPLC-ESI-QTOF-MS and multivariate analysis for use as a food ingredient. *Food Res. Int.* **2014**, *57*, 51–60. [CrossRef]
64. Fu, L.; Xu, B.T.; Xu, X.R.; Gan, R.Y.; Zhang, Y.; Xia, E.Q.; Li, H. Bin Antioxidant capacities and total phenolic contents of 62 fruits. *Food Chem.* **2011**, *129*, 345–350. [CrossRef]
65. Phenol-Explorer; USA Database Phenol-Explorer. Available online: http://phenol-explorer.eu/compounds (accessed on 18 February 2020).
66. Baret, P.; Septembre-Malaterre, A.; Rigoulet, M.; Lefebvre d'Hellencourt, C.; Priault, M.; Gonthier, M.-P.; Devin, A. Dietary polyphenols preconditioning protects 3T3-L1 preadipocytes from mitochondrial alterations induced by oxidative stress. *Int. J. Biochem. Cell Biol.* **2013**, *45*, 167–174. [CrossRef]
67. Asgary, S.; Sahebkar, A.; Afshani, M.R.; Keshvari, M.; Haghjooyjavanmard, S.; Rafieian-Kopaei, M. Clinical Evaluation of Blood Pressure Lowering, Endothelial Function Improving, Hypolipidemic and Anti-Inflammatory Effects of Pomegranate Juice in Hypertensive Subjects. *Phyther. Res.* **2014**, *28*, 193–199. [CrossRef] [PubMed]
68. Wu, S.B.; Meyer, R.S.; Whitaker, B.D.; Litt, A.; Kennelly, E.J. A new liquid chromatography-mass spectrometry-based strategy to integrate chemistry, morphology, and evolution of eggplant (*Solanum*) species. *J. Chromatogr. A* **2013**, *1314*, 154–172. [CrossRef] [PubMed]
69. Santiago-Silva, P.; Labanca, R.A.; Gloria, M.B.A. Functional potential of tropical fruits with respect to free bioactive amines. *Food Res. Int.* **2011**, *44*, 1264–1268. [CrossRef]
70. Bomtempo, L.L.; Costa, M.A.; Lima, H.; Engeseth, N.; Gloria, M.B.A. Bioactive amines in Passiflora are affected by species and fruit development. *Food Res. Int.* **2016**, *89*, 733–738. [CrossRef] [PubMed]

foods

Article

Willingness to Pay for Enhanced Mandatory Labelling of Genetically Modified Soybean Oil: Evidence from a Choice Experiment in China

Mingyang Zhang [1], Yubing Fan [2], Jingxia Cao [1], Lijun Chen [3] and Chao Chen [4,*]

[1] School of Business, Development Institute of Jiangbei New Area, Nanjing University of Information Science and Technology, 219 Niuliu Road, Nanjing 210044, China; 002668@nuist.edu.cn (M.Z.); 15162780938@163.com (J.C.)
[2] Texas A&M AgriLife Research, 11708 Highway 70 South, Vernon, TX 76384, USA; yubing.fan@ag.tamu.edu
[3] Food and Resource Economics Department, University of Florida, McCarty B 2115, Gainesville, FL 32611, USA; chenlij@missouri.edu
[4] College of Economics & Management, Nanjing Agricultural University, 1 Weigang, Nanjing 210095, China
* Correspondence: cchen@njau.edu.cn

Abstract: This study investigates consumers' preferences for mandatory labelling conveying the health and safety attributes of genetically modified soybean oil. The enhanced mandatory labelling includes allergen presence labelling, nutrient and compositional change labelling and traceability codes. The data were collected from a consumer survey in the eastern, central and western regions of China, with a total sample size of 804 respondents. We evaluated consumer willingness to pay (WTP) for enhanced mandatory labelling using a choice experiment approach. The results show that Chinese consumers are most favorable to traceability codes with a WTP of RMB 8.92, followed by allergen presences labelling, with RMB 6.57. Eastern consumers would like to pay a higher premium for the three types of enhanced mandatory labelling information, while central consumers only show a positive preference for traceability codes. The results imply that the efforts and policy strategies for enhanced mandatory labelling will benefit residents. Further studies can be expended to other genetically modified (GM) foods. This study provides information for the agency to improve mandatory GM food labelling management. This paper contributes to the growing body of the GM food literature by explicitly investigating consumer preference and WTP for mandatory labelling conveying the health and safety attributes of the GM foods.

Keywords: GM foods; food labelling; soybean oil; willingness to pay; choice experiment

Citation: Zhang, M.; Fan, Y.; Cao, J.; Chen, L.; Chen, C. Willingness to Pay for Enhanced Mandatory Labelling of Genetically Modified Soybean Oil: Evidence from a Choice Experiment in China. *Foods* **2021**, *10*, 736. https://doi.org/10.3390/foods10040736

Academic Editors: Adrián Rabadán and Rodolfo Bernabéu

Received: 17 February 2021
Accepted: 22 March 2021
Published: 30 March 2021

1. Introduction

As a fast developing country, China is the largest consumer and importer of genetically modified (GM) soybean [1]. The import of soybeans to China reached more than 95 million tons in 2017, which were used as processing materials [2]. Chinese consumers oppose GM foods due to their concerns about food safety [3]. Concerns of the potential risks of the GM foods are growing, particularly among civil society groups [4]. The lack of effective communication on GM foods leads to asymmetric information [5,6]. Mandatory labelling is considered a practical way to address the issue of asymmetric information in food labelling [7]. In order to disclose foods that are or may be bioengineered, the National Bioengineered Food Disclosure Law of U.S., passed by Congress in July 2016, directed the U.S. Department of Agriculture to establish the national mandatory standard. The National Bioengineered Food Disclosure Standard, which was announced in 2018, is simply a marketing label, and does not convey any information about the health, safety, or environmental attributes of the GM foods in comparison with their non-GM counterparts [8]. Similarly, China currently adopts qualitative mandatory labeling according to the catalogues, and the labelling catalogue includes 17 kinds of GM products within five categories, including soybeans, rape, corn, cotton, and tomato. As one of the directly processed GM agricultural products, the label of the GM soybean oil merely tells customers "the raw materials are GM soybeans." The usefulness of the simple GM food label is limited for consumers, because the current labelling only allows differentiating GM foods from non-GM food products [9]. Better information could increase consumers' support for GM foods, which further determines the implementation and success of the mandatory labelling policy [10]. However, mandatory labelling conveying the health and safety attributes for the improved transparency and openness of information inevitably leads to an increase in production cost, such as detecting cost. Therefore, this study investigates consumers' preferences for mandatory labelling conveying the health and safety attributes of the GM soybean oil.

Focusing on the developed countries, i.e., the US and European countries, much research has investigated consumer attitudes towards the mandatory labelling of the GM foods [11,12]. A research consensus has been reached that consumers may have greater demands for the mandatory labelling of the GM foods [11–14]. However, there are limited systematic studies on consumers' willingness to pay for the extra cost associated with the mandatory labelling information, especially on consumer preference and willingness to pay (WTP) for mandatory labelling conveying health and safety attributes. To the best of the authors' knowledge, no empirical study has been reported on WTP for the mandatory labelling of the GM foods in China.

Given the importance of mandatory labelling, this paper contributes to the literature by empirically analyzing urban consumers' preferences for mandatory labelling conveying health and safety attributes. Additional insights are provided for public service by evaluating consumers' WTP for the enhanced labelling of the GM soybean oil, including allergen presence labelling, nutrient and compositional change labelling, and traceability codes.

2. Literature Review

2.1. Debate on GM Food Labelling

Academia is divided on the pros and cons of the GM food labelling. On the one hand, some scholars holding a positive view believe that labelling can effectively tackle the problem of asymmetric information, which is of great benefit to forming a market with a remarkable separation of the GM and non-GM foods [14,15]. Mandatory labelling also contributes to GM food management, highlighting information such as place of origin, allergen presence, and detailed food ingredients, which are conducive to government regulation. Especially, the government can take timely and effective actions if any GM food safety problems occur [16,17]. The labelling policy has been found to be superior to an embargo in terms of consumer welfare and producer benefits [18]. In addition, GM labelling is closely related to consumers' right-to-know and assists consumers in making better informed purchase decisions [5].

On the other hand, other scholars holding a negative view believe that mandatory labelling obviously adds extra costs to the production and society, such as the adjustment cost, implementation cost, and monitoring cost [15,17]. For instance, upon the introduction of mandatory labelling in Europe and US, production costs increased by 17% and 6%, respectively [19]. If the GM food labels were added, each US household would pay an estimated USD 100 more on food every year [20].

2.2. Consumer Attitude towards GM Labelling

Relevant studies on consumer attitudes towards mandatory GM food labelling have reported findings in both developed and developing countries. In general, consumers have a strong preference for mandatory GM food labelling [11]. Marchant and Cardineau [21] analyzed the labelling debates in the US. Public opinion polls consistently show that 90 percent or more of Americans want foods to be labeled. Luck et al. [22] reported that over 80 percent of American consumers are supportive of implementing the mandatory labelling policy on GM products. Nep and O'Doherty [14] used data from a deliberative public engagement in British Columbia of Canada. In their survey, participants discussed the social and ethical implications of salmon genomics. The public called for mandatory labelling of transgenic salmon, and demanded labelling as a minimum requirement to allow consumers to choose whether to purchase GM foods. Participants showed strong distrust in the current supervision of the GM foods, and the perceived reluctance of biotechnology companies serves to fuel this distrust.

Further, much research has been conducted in developing countries. Huang and others [23] investigated 400 participants in Wuhan, China, and they found that more than four-fifths Chinese consumers demanded implementing mandatory GM food labelling policies. Deng et al. [24] found more than 90 percent of participants supported mandatory labelling, based on a survey of 260 participants from 11 provinces in China. Zhao et al. [25] investigated 1730 Chinese respondents' attitudes toward five different GM food labelling methods including no GM label, labels of meat fed by GM feeds, labels of cooking oil containing GM oil, labels of the GM condiments, and labels of non-GM ingredients. They found that those who were more familiar with genetically modified organisms (GMOs) or who trusted the government were more positive about GM labels. Kajale and Becker [13] conducted an interview among a sample of 298 students in India. They found that about 58 percent of college students supported mandatory GM food labelling, and about 44.63 percent believed the increased price should be jointly paid by consumer, producer and government.

2.3. Information Credibility and Adequacy of the GM Food Labels

U.S. consumers desire GM food labels to provide sufficient information relating to potential benefits and risks, which implies that the usefulness of a simple GM food label is limited for the public. As simple labels just allow consumers to differentiate GM food products from their non-GM counterparts, they do not include enough of the benefit and risk information that consumers desire to know [9]. Teisl and others [9] indicated that a simple GM label actually may not be beneficial to consumers who are anxious about GM contents but may be willing to accept the GM foods if the genetic modification provides any benefits. Moreover, excessive information on a GM food label may negatively affect consumers if they have limited knowledge of genetic engineering and GM foods [26]. Roe and Teisl [27] presented US consumers with some sample labels that contained different statements concerning the presence of the GM ingredients, and the consumers evaluated the credibility and adequacy of the information content. The result showed that a simple GM label just saying a product contains GM ingredients was considered more credible than the simple non-GM labels saying a product contains no GM ingredients. However, the consumers were more likely to judge the simple non-GM label as having provided an adequate amount of information for informed decisions to be made. They also found several significant improvements in the adequacy of simple GM labels when they mentioned the purpose of the GM usage, which significantly eroded the label's credibility rating. Hence, label credibility and label adequacy may remain opposite, but the provision of contact information may help resolve the credibility–adequacy trade-off.

2.4. Consumers' WTP for GM-Labeled Foods

Wolfe and others [11] found a significant premium for non-GM edamame even if there is no obvious difference between the overall sensory impression of the GM edamame and the non-GM counterparts in the US. This finding was similar to Huffman et al. [28] where a 14% premium was reported for non-GM vegetable oil, tortilla chips and potatoes compared to the GM-labeled counterparts. Likewise, Lusk et al. [29] found a premium of 25 cents per ounce for non-GM corn chips. Other scholars also reported similar findings and explained WTP for several kinds of the GM foods based on various functional GM foods (i.e., yield increasing, ripening controlling, protective, processed, nutrition improving GM foods), and crop classification (i.e., GM rice, GM vegetables, GM fruit, GM edible oil, etc.) [30–33]. However, there are limited systematic studies on the WTP for the extra cost associated with the mandatory labelling information, especially for the preference and WTP for mandatory labelling conveying health and safety attributes. In particular, no empirical research has been reported in China.

3. Choice Experiment

3.1. Identifying GM Labelling Policies

Currently, the GM organism labelling policies around the world fall into two types. One type is voluntary labelling, such as in Canada; the other is mandatory labelling, such as in the US, the EU, and China. In order to determine the exact rules applicable to labelling in the international and national context, we analyze the compilation of the codex committee on food labelling (CCFL), Canada's labelling policies, and the labelling legislation in the EU and its implementation in England. Since 1993, CCFL has begun to discuss the issue of the GM food labelling (1997 Text, 2001 Text, 2004 Text, 2008 Text, 2009 Text, 2010 Text and 2011 Text). Although it has yet to form a generally accepted international standard, the above text in the mandatory labelling has reached a consensus: there must be mandatory labelling in the presence of allergens [34]. Nevertheless, neither the regulations regarding the mandatory labelling of the GM food nor the provisions relating to the thresholds, exemptions and implementation are the same. The voluntary labelling model adopted in Canada requires labelling in the event of the presence of allergens or changes in the nutritional value or components [35]. The EU traceability and labelling regulation 1830/2003296 seeks to address the concerns about the lack of information to enable the labelling of the GM foods, and sets out the requirements for a document audit trail to account for and identify approved GM products throughout the marketing chain. This regulation summarizes the purpose: the traceability requirements for food and feed produced from GMOs should be established to facilitate the accurate labelling of such products. Its objective is to enable postmarket monitoring of health and the environment [35].

3.2. Experimental Design

A choice experiment (CE) approach was used to evaluate urban consumers' WTP for the attributes of enhanced mandatory labelling of the GM soybean oil. The CE model relies on random utility theory and factor value theory, and they indicate that the utility is from the attributes possessed by the item rather than item itself [36]. As for the enhanced mandatory GM food labelling, the combination of the labelling attributes and choice scenarios are formulated in the CE. Specifically, the consumer can obtain the utility v_k from the *k-th* labelling attribute, and the utility *V*, obtained from enhanced mandatory GM food labelling, equals to the sum of the utility v_k (k = 1, 2, . . . , K).

$$V = \lambda_1 v_1 + \lambda_2 v_2 + \ldots + \lambda_k v_k \tag{1}$$

where λ_k is the unknown parameter, referring to individual's preference for utility v_k. Consumer *i* must evaluate the utility U_{imn} from the enhanced mandatory GM food labelling associated with the alternative $m = 1, 2, \ldots, M$ in the *n-th* choice set. Within a given group of alternatives relating to a choice set, the consumer selects the utility-maximizing alternative. U_{imn} is a random variable that can be expressed as:

$$U_{imn} = V_{imn}\alpha + \mu_{mn} \tag{2}$$

where α_{mn} refers to the estimated parameter vector. μ_{mn} is the random disturbance term. The vector V_{imn} means sum of the utility obtained from the mandatory labelling attribute and payment vehicle associated with the alternative $m = 1, 2, \ldots, M$ in the *n-th* choice set [36].

This study adopts a choice experiment model on the GM soybean oil sales and labelling in China Including Regulations on Administration of Agricultural Genetically Modified Organisms Safety and Administrative Measures for Agricultural GMOs Labeling issued by the Ministry of Agriculture and Rural Affairs of P. R. China. This study follows the relevant literature [26,27,35,37–40], and the representative GM organism safety management policies. The CE model contained three labelling attributes and the payment vehicle (Table 1), that is, allergen presence labelling, nutrient and compositional change labelling, traceability codes, and price.

Table 1. Attributes and levels in the choice experiment.

Attributes	Levels	Description	Basis
(1). Allergens presence labelling	Disclosure, nondisclosure	Disclose the presence of allergens in GM soybeans oil	a, b, c
(2). Nutrient and compositional change labelling	Disclosure, nondisclosure	Disclose the changes in nutritional value or composition of the GM soybean oil, compared with non-GM counterpart.	b, c
(3). Traceability codes	Disclosure, nondisclosure	Traceability systems document the entire process of the GM soybean oil production. The systems allow for the separation of the GM soybean oil and non-GM products "from farm to fork," and serve the purpose of marketing and health protection.	d, e
(4). Price (RMB)	46, 53, 60	Price of a 5L jug of the GM soybean oil.	

a: Seven texts drafted by the Codex Alimentarius Commission (CAC) on Food Identification Subcommittee in the past decades. b: Food and Drugs Act, Food and Drug Regulations, and Consumer Packaging and Labeling Act by Canada; c: Draft Guidance for Industry: Voluntary Labeling Indicating Whether Foods Have or Have Not Been Developed Using Bioengineering by US; d: Traceability and Labelling Regulation and Regulation (EC) on Novel Foods and Novel Food Ingredients issued by EU; e: Genetically Modified Organisms (Traceability and Labelling) (England) Regulation 2004.312; RMB 6.80 = USD 1.

Each of the first three attributes include two levels (disclosure or nondisclosure), and the price includes three levels. Thus, there were 24 possible combinations in total. We can constitute 276 CE scenarios by pairing those combinations. After eliminating both the overlapping and theoretically contradictory CE scenarios, we conduct the screening experiment, and obtain twelve CE scenarios. These scenarios are randomly divided into two groups, with each contain six CE scenarios. A sample CE scenario is shown in (Figure 1).

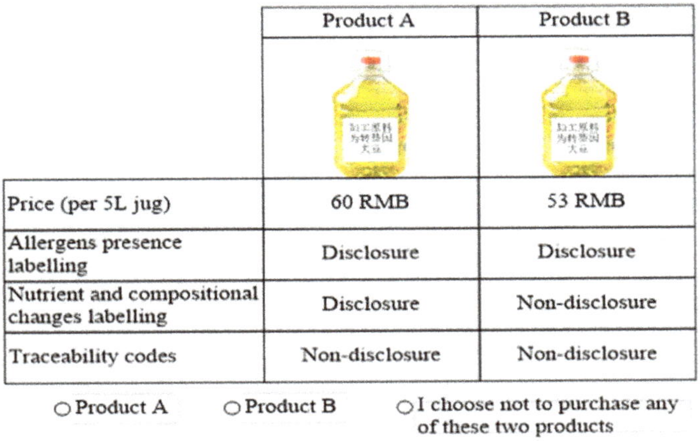

Figure 1. Example choice scenario.

In the experiment, GM soybean oil is selected as the analysis unit for four reasons. Firstly, the studies showed that Chinese consumers preferred to accept foods derived from bioengineered food rather than directly edible GM foods like GM soybean oil [41]. Therefore, it can be inferred that consumer demand for labelling information for the directly edible GM foods is the most urgent. Secondly, according to the GM organism safety certificates for both commercial planting and GM organisms imported as raw materials approved by the Ministry of Agriculture and Rural Affairs of China, currently there are only three kinds of directly edible GM foods on the Chinese market; that is, locally grown GM papaya, GM soybean oil and GM canola oil made from imported GM soybeans or rapeseed. The GM oil is labeled "The processing material is GM soybeans or rapeseeds." Thirdly, soybean oil is not only the most daily consumption edible oil in the majority of Chinese cities, but also the most popular with food processing enterprises and the catering industry. Fourthly, there are a wide range of alternatives to GM soybean oil and non-GM soybean oil available on the Chinese market, such as non-GM peanut oil, non-GM corn oil, non-GM sunflower oil, non-GM canola oil, non-GM rapeseed oil, and many kinds of oil blends. There are a variety of brands of edible oil in the Chinese market. Some only sell non-GM soybean oil (such as the Northeast soybean oil, Xinheshun and Qiansuihao, etc.), while some only sell GM soybean oil (such as Fortune, Jinlongyu, YuanBao, Fivelakes, etc.). Most enterprises produce only one or a few of the edible oils (such as only producing soybean oil, peanut oil, corn oil, sunflower oil, olive oil, rapeseed oil, or blended oil). In order to ensure the GM soybean oil, non-GM soybean oil and oil blends are identical in the brand, capacity and other aspects, this section uses "X" brand edible oil as the experimental unit, which is one of the top ten well-known brands of edible oil in China. Except for peanut oil, all other soybean oil substitutes are supplied in a 5-liter jug.

The prices of all kinds of "X" brand 5-liter edible oil are shown in (Table S1), which was presented to the respondents in the experiment. In this study, the price of 5L GM soybean oil (RMB 45.8, 1 USD=6.80 RMB) is set as the lower limit, and the price of 5L non-GM soybean oil (RMB 66.8) is set as the upper limit. According to the principle of isometric and rounding, the price is set at three levels: RMB 46, RMB 53, and RMB 60.

Additionally, the CE model follows a "randomized design" developed by Sawtooth Software, Inc. [42]. Compared to the fixed design, the randomized design can eliminate order and psychological context effects [43]. Additionally, the randomized designs are more efficient in asymmetric choice experiments when not all attributes have equal levels [44].

3.3. Mixed Logit Model

The mixed logit (ML) model is used to analyze the data collected in the choice experiment. The ML model (specified in the Equation (3) below) relaxes the independent of irrelevant alternative (IIA) assumption and allows individual variations in the attributes [45]. Meanwhile, the conditional logit model is typically used if the random terms follow independently identically distribution (IID) and assumes respondents having the same preference for the attributes. The likelihood ratio (LR) tests can be used to compare the two models [46]. If the null hypothesis that there is no difference between the two models is rejected, this indicates the ML model is more appropriate. In addition, a probability density function, $g(\alpha)$, is introduced for the coefficient of the presumed heterogeneous attributes. Namely, correlations between preferences are allowed and different respondents show different preferences for the attributes of enhanced mandatory labelling. The non-conditional probability P_{imn} of consumer i who chooses the m-th alternative in the n-th choice scenario can be get by calculating the integral of $g(\alpha)$ with respect to α.

$$P_{imn} = \int \frac{\exp(V_{imn}\alpha)}{\sum_{m=1}^{M} \exp(V_{imn}\alpha)} \tag{3}$$

The model assumes that $g(\alpha)$ functions of all nonpayment attributes follow normal distributions. The price attribute with a fixed coefficient equals the given market price and the other two reference prices are slightly higher than the market price. The parameter α refers to scaled marginal utility for a mandatory labeling attribute or price, due to scale normalization. Therefore, we can only interpret the relative magnitude of the other attributes and statistical significance by the parameter estimates. The WTP can be calculated from the negative marginal utility divided by the coefficient (α_p) of the price attribute [47]. Therefore, they are comparable across the results.

$$WTP = -\frac{\alpha_k}{\alpha_p} \tag{4}$$

4. Data

4.1. Survey Administration

The survey contained questions designed for the experiment and socio-demographic inquiries including gender, age, educational attainment, occupation, child, and income. The respondents were first provided with some detailed information on GM soybean imports and their connection to public interest. Each attribute was interpreted by the enumerators to make sure the respondents understood the survey. All the nonpayment attributes and price attributes in (Table 1) were shown to the respondents, who were asked whether they were willing to pay for the mandatory enhanced labelling of the GM soybean oil. They were also shown a sample of a CE scenario (see Figure 1) before the start of the experiment. We showed them what it would mean if "Product A" was chosen. The prices of all kinds of "X" brand 5-liter edible oil are shown in (Table S1), which were also presented to the respondents in the experiment. Two versions were developed to reflect the differences in the CE scenarios. Each respondent only took one version of the survey, assigned to six choice scenarios.

Adhering to the stratified random sampling, a self-administered questionnaire was utilized to collect data in the provincial capitals of the eastern, central and western regions. The eastern regions include Jinan of Shandong province, Nanjing of Jiangsu province, Shanghai, and Guangzhou of Guangdong province. The central regions include Changchun of Jilin province, Zhengzhou of Henan province, Hefei of Anhui province, Wuchang of Jiangxi province. The western regions include Lanzhou of Gansu province, and Guiyang of Guizhou province. This investigation was conducted at supermarkets and large-scale shopping malls by sixty-four undergraduates from Nanjing Agricultural University in 2017. Different social classes were sampled to avoid sample selection bias owing to sampling at a single site [48]. The survey enumerators approached potential respondents and invited them to participate if they wanted. The following steps were followed: (1) each respondent was confirmed to be an urban resident; (2) the selected respondents had food purchasing experience; (3) soybean oil was the family's main edible oil and was obtained through purchasing rather than through squeezing their own beans. After completing the survey, each respondent was offered a RMB 10 gift.

4.2. Descriptive Statistics

This analysis is based on 804 samples collected in the survey. The samples in the eastern, central and western regions account for 45.02%, 35.07% and 19.90%, respectively. Specially, there are 2172, 1688, and 960 choice scenarios in the eastern, central and western regions, respectively, because each respondent responds to six choice scenarios. A statistic summary of socio-demographics of the sampled urban consumers is shown in (Table 2). Compared with the population, i.e., the national urban and rural residents, from the 2017 China statistical yearbook, the sample includes fewer males and shows a better education attainment, with 63.18% attending a professional school or holding a college or higher degree. The sample includes more young people under 45 years old. In addition, about 8% of the respondents have a job relating to biotechnology, and 56.47% of the families have minors. The average monthly household disposable income roughly follows a normal distribution, with the categories RMB 4001–6000, RMB 6001–8000, and RMB 8001–10,000 accounting for 31.84%, 16.92%, and 12.69%, respectively.

Table 2. Definition and descriptive statistics of the demographic variables.

Variable	Description	Sample Mean [a]	Population Mean [b]
Gender	0 = male; 1 = female.	0.500 (0.500)	0.499
Age	0 = young people (18–44); 1 = middle-aged or senior people (≥45).	0.254 (0.435)	
Education attainment	0 = senior high school or below; 1 = professional school, college degree or above.	0.632 (0.483)	0.401
Occupation	Whether your work is related to biotechnology? (0 = no; 1 = yes)	0.079 (0.271)	
Child	Whether your family has minors (≤15)? (0 = no; 1 = yes)	0.435 (0.496)	
Income	Monthly household disposable income (1 = RMB RMB 2001–4000; 3 = RMB 4001–6000; 4 = RMB 6001–8000; 5 = RMB 8001–10,000; 6 = RMB 10,001–12,000; 7 = RMB 12,000 or above)	3.398 (1.537)	

[a] Standard deviation in the parentheses. [b] The population includes urban and rural residents based on information of 31 provinces from "2017 China Statistical Yearbook". [b] RMB 6.80 = USD 1.

Table 3 shows a statistical summary of the variables used in the ML regressions. The consumers who are willing to pay for the enhanced mandatory labelling of the GM soybean oil account for 67.40%, 58.90%, and 57.50% in the eastern, central and western regions, respectively. From the nationwide perspective, the means of the three attributes are −0.087, −0.044, and −0.011, respectively. The average of the prices is RMB 35.21.

Table 3. Definition and descriptive statistics of the variables used in mixed logit model.

Variable	Description	Nationwide Consumers (N[a] = 14,472)	Eastern Consumers (N[a] = 6516)	Central Consumers (N[a] = 5076)	Western Consumers (N[a] = 2880)
Stated intention	0 = unwilling to pay; 1 = willing to pay	0.624 (0.484)	0.674 (0.469)	0.589 (0.492)	0.575 (0.494)
Whether to choose the option	0 = no; 1 = yes	0.333 (0.471)	0.333 (0.471)	0.333 (0.471)	0.333 (0.471)
Allergen presence labelling	0 = no; 1 = yes; −1 = I do not choose either option A or B	−0.087 (0.757)	−0.095 (0.751)	−0.082 (0.760)	−0.076 (0.765)
Nutrient and compositional change labelling		−0.044 (0.788)	−0.084 (0.759)	−0.023 (0.802)	0.007 (0.821)
Traceability codes		−0.111 (0.737)	−0.111 (0.737)	−0.111 (0.737)	−0.111 (0.737)
Price	RMB 46; RMB 53; RMB 60; 0 if option C is chosen	35.210 (25.360)	35.370 (25.480)	35.120 (25.290)	34.990 (25.200)

[a] N means the number of options. Standard errors in the parentheses.

5. Estimation Results

We conducted the mixed logit regressions using the simulated maximum likelihood estimator. Firstly, the correlation test shows that there is correlation between each pair of the attribute variables. For example, the correlation coefficient between allergen presence labelling and nutrient and compositional change labelling is 0.74. Obviously, the correlation test result is in conflict with the IID assumption of the conditional logit model. Therefore, we run the ML model with correlated normally distributed coefficients. The result of the LR test for the nationwide sample is 3858 and significant at $p < 0.001$, which indicates that the null hypothesis is rejected. In other words, the respondents have heterogeneous preferences. Therefore, the conditional logit model has a poorer fit compared to the ML model.

Table 4 provides the estimation results of respondents' preferences for the enhanced mandatory labelling of the GM soybean oil. The results show that price has a negative effect in all equations, indicating an increase in GM soybean oil price decreases the probability of a consumer choosing the oil. More importantly, most proposed enhanced GM mandatory labelling information is positive.

Table 4. Estimation results of the mixed logit model.

	Nationwide Consumers		Eastern Consumers		Central Consumers		Western Consumers	
	All Samples	**Stated Intention = 1**	**All Samples**	**Stated Intention = 1**	**All Samples**	**Stated Intention = 1**	**All Samples**	**Stated Intention = 1**
Allergen presence labelling	0.427 ***	1.074 ***	0.725 ***	1.359 ***	0.286	0.994 ***	0.115	0.528 ***
	(0.112)	(0.081)	(0.170)	(0.130)	(0.189)	(0.135)	(0.237)	(0.172)
Nutrient and compositional change labelling	0.104	0.502 ***	0.530 ***	0.754 ***	−0.181	0.404 ***	−0.242	0.015
	(0.101)	(0.067)	(0.138)	(0.103)	(0.187)	(0.121)	(0.230)	(0.145)
Traceability codes	0.579 ***	1.118 ***	0.919 ***	1.382 ***	0.411 **	1.010 ***	−0.094	0.744 ***
	(0.115)	(0.092)	(0.166)	(0.142)	(0.202)	(0.156)	(0.320)	(0.213)
Price	−0.065 ***	−0.061 ***	−0.082 ***	−0.079 ***	−0.055***	−0.054 ***	−0.040 ***	−0.030 ***
	(0.005)	(0.005)	(0.007)	(0.007)	(0.008)	(0.008)	(0.011)	(0.010)
Coefficient covariance								
v11	1.314 ***	0.682 ***	1.471 ***	0.893 ***	−1.230 ***	0.502 ***	0.988 ***	0.591 ***
	(0.111)	(0.085)	(0.172)	(0.125)	(0.175)	(0.111)	(0.201)	(0.133)
v 21	0.696 ***	0.200	0.419 **	−0.093	−1.055 ***	0.210	0.815 ***	0.688 ***
	(0.127)	(0.140)	(0.207)	(0.179)	(0.219)	(0.242)	(0.251)	(0.140)
v 31	0.852 ***	0.194	0.914 ***	0.366 **	−0.999 ***	0.499 **	0.336	−0.958 ***
	(0.120)	(0.121)	(0.157)	(0.145)	(0.208)	(0.194)	(0.308)	(0.269)
v 22	1.031 ***	0.911 ***	1.054 ***	0.983 ***	1.029 ***	1.055 ***	−0.410	−0.032
	(0.096)	(0.095)	(0.173)	(0.133)	(0.172)	(0.148)	(0.272)	(0.379)
v 32	−0.574 ***	−0.666 ***	−0.509 ***	−0.470 ***	−0.992 ***	−0.804 ***	0.944 **	−0.584
	(0.117)	(0.115)	(0.192)	(0.141)	(0.163)	(0.168)	(0.369)	(0.470)
v 33	0.984 ***	0.683 ***	0.865 ***	0.671 ***	0.406	0.053	1.655 ***	0.758 **
	(0.130)	(0.124)	(0.172)	(0.142)	(0.287)	(0.422)	(0.289)	(0.303)
Log likelihood	−5118.33	−4397.01	−2353.36	−2054.82	−1741.35	−1477.03	−990.21	−831.17
LR chi2(6)	3858.12 ***	835.34 ***	1739.81 ***	464.86 ***	1352.73 ***	241.50 ***	717.31 ***	135.14 ***
P value	<0.001	<0.001	<0.001	<0.001	<0.001	<0.001	<0.001	<0.001
Number of options	14472	9036	6516	4392	5076	2988	2880	1656

The parameters in the bottom panel of the output are the elements of the lower-triangular matrix L. Standard errors in the parentheses. *** and ** denote statistical significance at the 0.01, 0.05 and 0.10 levels, respectively.

The coefficients of the three variables of labelling attributes are 0.427 ($p < 0.01$), 0.104, and 0.579 ($p < 0.01$) on a national scale (Table 4). The results suggest that Chinese urban consumers are in favor of the enhanced mandatory labelling of the GM soybean oil. The most attractive and influential labelling attribute is traceability codes, followed by allergen presence labelling with a smaller effect. The least important labelling is nutrient and compositional change labelling, and consumers are more likely to select an alternative based on other enhanced labelling included.

To better understand consumers' preference for mandatory labelling conveying health and safety attributes, we estimate the WTP values using the parameter estimates from the ML model. The WTP values for each attribute are shown in Table 5. The magnitude of WTP and their ranks are consistent with that of the coefficient estimates from the ML model (Table 4). Positive WTP values represent the amount of money that the consumers are willing to pay for the specific labelling attributes. The highest WTP value is found for the traceability codes. Specifically, the consumers are more likely to pay for the traceability codes nationwide, with a payment of 8.92 RMB, followed by the allergen presence labelling with a value of 6.57 RMB. Regionally, eastern consumers show a positive preference for all three attributes, with the payment amounts of 11.24 RMB, 8.87 RMB, and 6.48 RMB for traceability codes, allergen presence labelling, and nutrient and compositional change labelling, respectively. Central consumers only show a positive preference for the traceability codes, i.e., 7.41 RMB. However, western consumers show no preference.

Table 5. Willingness to pay for a premium for enhanced mandatory labeling information of the GM soybean oil.

Attribute	Nationwide Consumers		Eastern Consumers		Central Consumers		Western Consumers	
	All Samples	Stated Intention = 1	All Samples	Stated Intention = 1	All Samples	Stated Intention = 1	All Samples	Stated Intention = 1
Allergen presence labelling	6.57 ***	17.50 ***	8.87 ***	17.23 ***	5.15	18.28 ***	2.85	17.58 ***
	(3.72, 9.42)	(15.85, 19.15)	(5.66, 12.09)	(15.18, 19.28)	(−0.59, 10.89)	(15.17, 21.38)	(−7.82, 13.52)	(10.12, 25.03)
Nutrient and compositional change labelling	1.60	8.17 ***	6.48 ***	9.56 ***	−3.27	7.42 ***	−5.99	0.51
	(−1.34, 4.54)	(6.35, 10.00)	(3.42, 9.55)	(7.20, 11.93)	(−10.36, 3.83)	(3.73, 11.10)	(−18.96, 6.98)	(−8.73, 9.76)
Traceability codes	8.92 ***	18.22 ***	11.24 ***	17.52 ***	7.41 ***	18.57 ***	−2.33	24.76 ***
	(5.98, 11.86)	(16.34, 20.09)	(8.20, 14.28)	(15.53, 19.52)	(1.17, 13.63)	(14.88, 22.26)	(−18.37, 13.71)	(11.91, 37.61)

Lower bound and upper bound for 95% confidence interval in the parentheses. *** denote statistical significance at the 0.01, 0.05 and 0.10 levels, respectively.

Furthermore, the survey results show that about 62.44% of the urban consumers state that they are willing to pay for the enhanced mandatory labelling (Table 3). For those who are willing to pay, the average WTP is RMB 18.22 for traceability codes, followed by RMB 17.50 for allergen presence labelling. The WTP for nutrient and compositional change labelling is the smallest with a payment amount of RMB 8.17.

6. Discussion

Chinese urban consumers show a positive preference for mandatory labelling conveying some information about the safety attributes of GM foods. This is largely because a simple GM food label, such as "the raw material is GM soybeans," only allows differentiating GM foods from their non-GM counterparts. Most consumers would like to see more detailed information about the potential benefits and risks on GM food labels [9]. Urban consumers are more likely to pay for traceability codes, followed by allergen presence labelling, while nutrient and compositional change labelling is least important. Our results are consistent with the findings of Roe and Teisl [27] who suggested that providing contact sources that consumers can use to obtain more information could resolve the credibility–adequacy trade-off. They also proposed several improvements in the adequacy of simple GM labels, such as adding the purpose of the GM ingredients' usage. While this addition on the label also greatly erodes the label's credibility and retains the opposition of label credibility and adequacy [27].

Our results show moderate regional heterogeneity in the preference and WTP. Among those who are willing to pay, the western consumers show a strong preference for allergen presence labelling and traceability codes with the values of RMB 17.58 and RMB 24.76, respectively, while central consumers also have a stronger preference for both allergen presence labelling and traceability codes, and the WTP values are RMB 18.28 and RMB 18.57, which are higher than the eastern levels. On average, the per capita disposable income in the eastern cities is higher than that in the central cities, which is turn higher than that in the western cities [49]. Meanwhile, the consumers in the higher income region would not like to pay a premium for the enhanced mandatory labelling of the GM soybean oil. Additionally, we found that compared to the respondents who are unwilling to pay, the respondents who are willing to pay have a lower per capita disposable income. This is consistent with the findings of Wolfe and others [11] who reported that the urban households with a higher income can afford the non-GM oil. In general, the higher income households are more cautious about food choice [50,51]. Thus, they are more likely to purchase the non-GM oil rather than being willing to pay for the enhanced labelling of the GM foods. For lower income consumers, GM oil may be their main edible oil because of the low price. They would like to know more about GM foods, and have a stronger demand for right-to-know. Mandatory labelling may be a practical way to address the issue of asymmetric information in food labelling [7]. Hence, consumers from western China would like to pay a higher premium for the enhanced mandatory labelling of the GM soybean oil.

On average, the eastern consumers have a higher education level than those from the central region, whose education level is in turn higher than that of consumers from the western region. Educational attainment may determine food preference; therein lies a useful pointer for the policy makers [52]. Well-educated consumers may be more concerned about GM foods, because they may be worried about the uncertainty of transgenic technology, but they may be not in fact be aware of GM products [53,54].

7. Conclusions

This paper contributes to the growing body of the GM food literature by explicitly investigating consumer preference and WTP for mandatory labelling conveying the health and safety attributes of the GM foods. The results signify that consumers recognize the importance of investing in the mandatory labelling conveying safety information. This suggests the efforts and policy strategies for enhanced mandatory labelling will benefit Chinese citizens. This is encouraging because financial and technical assistance from the government can target certain interest groups, rather than distributing the resources to satisfy all groups. It may be more interesting to agency leaders to consider the specific WTP amounts for the three types of enhanced mandatory labelling information. Allergen presence labelling and nutrient and compositional change labelling can better help consumers understand the potential risks and benefits of the GM foods, but neither is highly ranked in terms of the WTP values. Instead, the traceability codes show the highest WTP value. This is interesting because the traceability codes may help consumers know where the products come from, but it would not inform them of the potential risks and benefits. Conversely, it may be the nature of the right-to-know of traceability codes that makes them more valuable to the public. While government agencies are responsible for improving mandatory GM food labelling management for the benefit of the public, it is critical to include publicly linked policies, such as consumer WTP for enhanced labelling, to gain more support from the public.

Policies encouraging consumers to make purchase decisions that match personal preference are inherently desirable, regardless of the end-user characteristics or process attributes. These policies should be cost-effective. Unfortunately, our results do not present the costs or benefits of instituting an enhanced mandatory labelling program. A policy decision to impose enhanced mandatory labelling should recognize both its benefits and costs, while considering whether the practitioners are equipped or facilitated to implement the policy. The research does not conclude that an enhanced mandatory labelling program should be instituted. Rather, the findings provide guidance on how an enhanced mandatory labelling program should look like if such a program is warranted. Nevertheless, further research is needed, including calculating the additional costs and evaluating the benefits. Additionally, this paper is restricted to GM soybean oil, while future research can expand this approach to other GM foods.

Supplementary Materials: The following are available online at https://www.mdpi.com/article/10.3390/foods10040736/s1, Table S1. Prices of the "X" brand edible oil in the same size.

Author Contributions: Conceptualization, M.Z., Y.F. and C.C.; Data curation, M.Z.; Formal analysis, Y.F.; Funding acquisition, M.Z.; Investigation, M.Z., J.C. and C.C.; Methodology, M.Z., Y.F., J.C. and L.C.; Project administration, C.C.; Resources, C.C.; Software, J.C. and L.C.; Writing—original draft, M.Z., Y.F., J.C. and L.C.; Writing—review & editing, C.C. All authors have read and agreed to the published version of the manuscript.

Funding: This research was funded by the National Natural Science Foundation of China (71603129; 71673139).

Institutional Review Board Statement: Not applicable.

Informed Consent Statement: Informed consent was obtained from all subjects involved in the study.

Data Availability Statement:. Not applicable.

Conflicts of Interest: No conflict of interest exits in this manuscript, and it has been approved by all authors for publication.

References

1. Pray, C.E.; Huang, J.; Hu, R.; Deng, H.; Yang, J.; Morin, X.K. Prospects for cultivation of genetically engineered food crops in China. *Glob. Food Secur.* **2018**, *16*, 133–137. [CrossRef]
2. Yu, C.; Deng, H.; Hu, R. Attitude gaps with respect to GM non-food crops and GM food crops and confidence in the Government's management of biotechnology: Evidence from Beijing consumers, Chinese farmers, journalists, and government officials. *Sustainability* **2020**, *12*, 324. [CrossRef]
3. Zhang, Y.; Jing, L.; Bai, Q.; Shao, W.; Feng, Y.; Yin, S.; Zhang, M. Application of an integrated framework to examine Chinese consumers' purchase intention toward genetically modified food. *Food Qual. Pref.* **2018**, *65*, 118–128. [CrossRef]
4. Oh, J.; Ezezika, O.C. To label or not to label: Balancing the risks, benefits and costs of mandatory labelling of the GM food in Africa. *Agric. Food Secur.* **2014**, *3*, 8. [CrossRef]
5. Vecchione, M.; Feldman, C.; Wunderlich, S. Consumer knowledge and attitudes about genetically modified food products and labelling policy. *Int. J. Food Sci. Nutr.* **2015**, *66*, 329–335. [CrossRef] [PubMed]
6. Ribeiro, T.G.; Barone, B.; Behrens, J.H. Genetically modified foods and their social representation. *Food Res. Int.* **2016**, *84*, 120–127. [CrossRef]
7. Roe, B.; Teisl, M.F.; Deans, C. The economics of voluntary versus mandatory labels. *Annu. Rev. Resour. Econ.* **2014**, *6*, 407–427. [CrossRef]
8. USDA Agricultural Marketing Service. BE Disclosure. Available online: https://www.ams.usda.gov/rules-regulations/be (accessed on 1 January 2019).
9. Teisl, M.F.; Garner, L.; Roe, B.; Vayda, M. Labeling genetically modified foods: How do US consumers want to see it done? *Agbioforum* **2003**, *6*, 1522–1936X.
10. Bain, C.; Dandachi, T. Governing GMOs: The (counter) movement for mandatory and voluntary non-GMO labels. *Sustainability* **2014**, *6*, 9456–9476. [CrossRef]
11. Wolfe, E.; Popp, M.; Bazzani, C.; Nayga, R., Jr.; Danforth, D.; Popp, J.; Chen, P.; Seo, H. Consumers' willingness to pay for edamame with a genetically modified label. *Agribusiness* **2018**, *34*, 283–299. [CrossRef]
12. Hallman, W.K.; Cuite, C.L.; Morin, X. Public perceptions of labeling genetically modified foods. Available online: http://humeco.rutgers.edu/documents_PDF/news/GMlabelingperceptions.pdf (accessed on 1 November 2018).
13. Kajale, D.B.; Becker, C.T. Determinants of consumer support for mandatory labeling of genetically modified food in India: A student survey. *Br. Food J.* **2013**, *115*, 1597–1611. [CrossRef]
14. Nep, S.; O'Doherty, K. Understanding public calls for labeling of genetically modified foods: Analysis of a public deliberation on genetically modified salmon. *Soc. Nat. Resour.* **2013**, *26*, 506–521. [CrossRef]
15. Bansal, S.; Gruère, G.P. Implications of mandatory labeling of the GM food in India: Evidence from the supply side. *Food Policy* **2012**, *37*, 467–472. [CrossRef]
16. Dannenberg, A.; Scatasta, S.; Sturm, B. Mandatory versus voluntary labelling of genetically modified food: Evidence from an economic experiment. *Agric. Econ.* **2011**, *42*, 373–386. [CrossRef]
17. Zainol, Z.A.; Nordin, R.; Akpoviri, F.I. Mandatory labelling of genetically modified (GM) foods. *Int. Environ. Agreem. Politics Law Econ.* **2015**, *15*, 199–216. [CrossRef]
18. Hobbs, J.E.; Kerr, W.A. Consumer information, labelling and international trade in agri-food products. *Food Policy* **2006**, *31*, 78–89. [CrossRef]
19. Zhao, J.; Sheng, J. Problem and confusion about GMO food labeling. *J. China Agric. Univ.* **2015**, *20*, 1–8.
20. Gao, W.; Luo, Y. Studies on the pros and cons on global GM food labeling debate. *J. Chin. Inst. Food Sci. Technol.* **2016**, *1*, 1–9. (In Chinese)
21. Marchant, G.E.; Cardineau, G.A. The labeling debate in the United States. *GM Crops Food* **2013**, *4*, 126–134. [CrossRef] [PubMed]
22. Lusk, J.L.; Mcfadden, B.R.; Wilson, N. Do consumers care how a genetically engineered food was created or who created it? *Food Policy* **2018**, *78*, 81–90. [CrossRef]
23. Huang, J.; Ji, Z.; Feng, L.; Zhang, D. Impact of labeling management system on consumers' purchase intention to GMF: Takes Wuhan as example. *J. China Agric. Univ.* **2013**, *18*, 220–225. (In Chinese)
24. Deng, X.; Guo, Y.; Su, H. The relationship between labeling and cognition degree of genetically modified food. *Forum Sci. Technol. China* **2018**, *272*, 170–178. (In Chinese)
25. Zhao, Y.; Hu, R.; Deng, H. Awareness and attitude toward GM labeling: Evidence from China. In Proceedings of the 30th International Conference of Agricultural Economists, Vancouver, BC, Canada, 28 July–2 August 2018.
26. O'Fallon, M.; Gursoy, D.; Swanger, N. To buy or not to buy: Impact of labeling on purchasing intentions of genetically modified foods. *Int. J. Hosp. Manag.* **2007**, *26*, 117–130. [CrossRef]
27. Roe, B.; Teisl, M. Genetically modified food labeling: The impacts of message and messenger on consumer perceptions of labels and products. *Food Policy* **2007**, *32*, 49–66. [CrossRef]

28. Huffman, W.E.; Shogren, J.F.; Rousu, M.; Tegene, A. Consumer willingness to pay for genetically modified food labels in a market with diverse information: Evidence from experimental auctions. *J. Agric. Resour. Econ.* **2003**, *17*, 481–502.
29. Lusk, J.L.; Daniel, M.S.; Mark, D.R.; Lusk, C.L. Alternative calibration and auction institutions for predicting consumer willingness to pay for nongenetically modified corn chips. *J. Agric. Resour. Econ.* **2001**, *15*, 40–57.
30. Puduri, V.S.; Govindasamy, R.; Vellangany, I. Willingness to buy genetically modified foods: An analysis of plant and animal origins in the US. *IUP J. Agric. Econ.* **2011**, *8*, 47–63.
31. De Steur, H.; Buysse, J.; Feng, S.; Gellynck, X. Role of information on consumers' willingness-to-pay for genetically-modified rice with health benefits: An application to China. *Asian Econ. J.* **2013**, *27*, 391–408. [CrossRef]
32. Thorne, F.; Fox, J.A.; Mullins, E.; Wallace, M. Consumer willingness-to-pay for genetically modified potatoes in Ireland: An experimental auction approach. *Agribusiness* **2017**, *33*, 43–55. [CrossRef]
33. De Steura, H.; Blancquaertb, D.; Strobbeb, S.; Fengc, S.; Buyssea, J.; Stoved, C.; Lambertd, W.; Van Der Straetenb, D.; Gellyncka, X. Consumer acceptance and willingness-to-pay for genetically modified foods with enhanced vitamin levels. In *Genetically Modified Organisms in Food: Production, Safety, Regulation and Public Health*; Elsevier Academic Press: Amsterdam, Netherlands, 2016; pp. 195–206.
34. Codex-Committee-on-Food-Labelling, proposed draft compilation of codex texts relevant to labelling of foods derived from modern biotechnology. In *Codex Alimentarius Commission*; Quebec, QC, Canada, 2011; p. 43.
35. Sowmya-Latha, H.V. *The Codex: Labelling of Food Derived from Modern Biotechnology. The Degree of Master of Laws*; Western University: London, ON, Canada, 2012; p. 747. Available online: https://ir.lib.uwo.ca/etd/747 (accessed on 1 March 2017).
36. Mcfadden, D. Conditional logit analysis of qualitative choice behavior. *Front. Econom.* **1974**, *7*, 105–142.
37. Fu, W. On international standards of genetically modified food labeling: From perspective of codex alimentarius commission. *J. Huazhong Agric. Univ. Soc. Sci. Ed.* **2012**, *5*, 82–86. (In Chinese)
38. Zhou, C. Improvement of China 's genetically modified food labeling system under the framework of international. *Law Seek.* **2016**, *6*, 53–57.
39. Fu, W. Legislative cracking of the plight of genetically modified food labeling in China. *J. Zhongzhou* **2015**, *9*, 55–61. (In Chinese)
40. Miraglia, M.; Berdal, K.G.; Brera, C.; Corbisier, P.; Holstjensen, A.; Kok, E.J.; Marvin, H.J.; Schimmel, H.; Rentsch, J.; Van, J.R. Detection and traceability of genetically modified organisms in the food production chain. *Food Chem. Toxicol.* **2004**, *42*, 1157–1180. [CrossRef]
41. Zhang, M. *Influence of Source Credibility of Genetically Modified on Consumer Attitudes: The Mechanism and Empirical Research*; Nanjing Agricultural University Press: Nanjing, China, 2015.
42. Pinnell, J. Comment on Huber: Practical suggestions for CBC studies. In Proceedings of the Sawtooth Software Conference, Sequim, WA, USA, 6–8 October 2004.
43. Day, B.; Bateman, I.J.; Carson, R.T.; Dupont, D.; Louviere, J.J.; Morimoto, S.; Scarpa, R.; Wang, P. Ordering effects and choice set awareness in repeat-response stated preference studies. *J. Environ. Econ. Manag.* **2012**, *63*, 73–91. [CrossRef]
44. Mulhern, M.G. Assessing the relative efficiency of fixed and randomized experimental designs. In Proceedings of the Sawtooth Software Conference, Sequim, WA, USA, 2 February 1999.
45. Revelt, D.; Train, K. Mixed logit with repeated choices: Households' choices of appliance efficiency level. *Rev. Econ. Stat.* **1998**, *80*, 647–657. [CrossRef]
46. Penn, J.; Hu, W.; Cox, L.; Kozloff, L. Resident and tourist preferences for stormwater management strategies in Oahu, Hawaii. *Ocean Coast. Manag.* **2014**, *98*, 79–85. [CrossRef]
47. Adamowicz, W.; Louviere, J.; Williams, M. Combining revealed and stated preference methods for valuing environmental amenities. *J. Environ. Econ. Manag.* **1994**, *26*, 271–292. [CrossRef]
48. Kornelis, M.; De, J.J.; Frewer, L.; Dagevos, H. Consumer selection of food-safety information sources. *Risk Anal.* **2007**, *27*, 327–335. [CrossRef] [PubMed]
49. Menggen, C. Price difference, income inequality and regional penn effect. *Econ. Perspect.* **2020**, *710*, 51–65.
50. Zhang, M.; Fan, Y.; Chen, C.; Cao, J.; Pu, H. Consumer perception, mandatory labeling, and traceability of the GM soybean oil: Evidence from Chinese urban consumers. *GM Crops Food* **2021**, *12*, 36–46. [CrossRef] [PubMed]
51. French, S.A.; Tangney, C.; Crane, M.M.; Wang, Y.; Appelhans, B.M. Nutrition quality of food purchases varies by household income: The SHoPPER study. *BMC Public Health* **2019**, *19*, 231. [CrossRef] [PubMed]
52. Caillavet, F.; Darmon, N.; Létoile, F.; Nichèle, V. Four decades of household food purchases: Changes in inequalities of nutritional quality in France, 1971–2010. *Econ. Stat.* **2019**, *513*, 69–89.
53. Zhao, Y.; Deng, H.; Yu, C.; Hu, R. The Chinese public's awareness and attitudes toward genetically modified foods with different labeling. *NPJ Sci. Food* **2019**. [CrossRef] [PubMed]
54. Xin, M. Cognitive attainment and acceptance intention of Henan consumers on genetically modified food. *China Soft Sci.* **2017**, *9*, 60–68.

foods

MDPI

Article

Innovation Strategies of the Spanish Agri-Food Sector in Response to the Black Swan COVID-19 Pandemic

Margarita Brugarolas [1,*], Laura Martínez-Carrasco [1], Adrián Rabadán [2] and Rodolfo Bernabéu [2]

1 Escuela Politécnica Superior de Orihuela (EPSO), Universidad Miguel Hernández, Avenida de la Universidad de Elche s/n, 03202 Elche, Spain; lmartinez@umh.es
2 Escuela Técnica Superior de Ingenieros Agrónomos y de Montes (ETSIAM), Universidad de Castilla-La Mancha, Campus Universitario s/n, 02071 Albacete, Spain; adrian.rabadan@uclm.es (A.R.); Rodolfo.Bernabeu@uclm.es (R.B.)
* Correspondence: mbrugaro@umh.es

Received: 30 October 2020; Accepted: 4 December 2020; Published: 8 December 2020

Abstract: Health, financial, and social crises cause variations in the buying behaviour of food consumers as well as in the value they assign to food attributes and the place of purchase, leading to consumers with profiles that are more susceptible to these changes than others. Thus, it was observed that 61.4% of consumers modified their buying behaviour at the onset of the COVID-19 pandemic, with those who modified it the most being the people who stockpiled the most food and went panic buying more often. This has made it possible to establish the profile of different significant consumer segments, and as a response, food production/distribution companies can implement different innovative strategies aimed at decreasing the impact of stockpiling and, therefore, the shortage of food. The possible strategies that companies can put into effect are creating a stock of non-perishable foods, increasing production capabilities in a sustainable way and, especially in light of the results obtained, boost the online sale and distribution of foods, with the goal of decreasing the amount of people in shops (which decreases the spreading of the pandemic and favours health) and preventing consumers from observing possible circumstantial shortages that would only encourage stockpiling and panic buying, even among consumers who have not changed their buying behaviour.

Keywords: agri-food marketing; consumer behaviour; economic and social crisis; health; innovation

1. Introduction

Black swan is a term coined by investor Nassim Nicholas Taleb in 2007 and it has been used to designate unpredictable events with a high economic and social impact [1]. Some researchers have defined the current health crisis as a black swan [2,3]. However, others, Taleb included, consider that this pandemic situation could have been predicted. In fact, experts in infectious disease/public health protection have been warning us for decades that a global pandemic involving a highly infectious respiratory disease virus was a plausible scenario [4–6].

There have been noteworthy precedents for other pandemics triggered by viruses or bacteria with a disastrous effect on human history [7–11]. Despite that, humanity has once again shown that it was not prepared to tackle these situations, even though being prepared for disasters can minimize damage to our health, lives, and property [12].

The current outbreak has been caused by severe acute respiratory syndrome coronavirus 2 (SARSCOV-2). Apart from the severe health problems [13], this pandemic has had a ripple effect on every aspect of human life as we know it, shaking our current society and affecting social, political and economic issues in a terrible way [3].

In this context, governments around the world have issued unprecedented policies and guidelines to save lives by reducing the pace and extent of COVID-19 infections ("flatten the curve") [14,15].

Lockdown has been one of the most common measures in the beginning of the pandemic in a large number of countries, and it is considered the best public health containment strategy available [16]. Without exception, national lockdowns throughout the world have caused considerable disruption to individuals, families, households, communities, national economies, and societies as a whole [17,18] and in most countries, governments announcing lockdowns has led to the panic buying of food as well as premature shortages of goods and services [3,19–21].

Food is a basic need, and preserving the supply is essential during crises. The identification of the most influential variables on consumer behaviour is essential for companies in order to satisfy the demands of increasingly sophisticated and demanding food consumers [22]. Until now, different researchers developed models to explain consumer behaviour in the food selection process [23–25], which have been widely implemented.

Nevertheless, despite the common patterns of consumption being well known, the extraordinary situation that mankind has found itself immersed in is not conventional at all. Classic behavioural theories explaining consumer behaviour may not apply in this context [26].

The goal of this study was to analyse the food buying changes that have taken place in Spain in two time periods, with the reference point separating them being the declaration of state of alarm on 14 March 2020 due to the health crisis caused by COVID-19. The first time period is the week just before the state of alarm was declared, and the second is the first week after said declaration (lockdown), in order to compare them to regular buying behaviours.

In addition, the study also sought to learn the profile of food consumers, identifying those who are more sensitive to stressful situations and will change their regular buying behaviour the most, therefore being more susceptible to go panic buying.

With knowledge of the buying behaviour and profile of consumers in black swan situations (such as the COVID-19 pandemic, for example), agri-food companies will be able to plan ahead with business innovation strategies that allow them to, on one hand, decrease the troublesome behaviours of consumers, and on the other, guarantee the market supply in foreseeable economic, social, and health crises.

1.1. Panic Buying

Faced with the "fear of the unknown", consumers took precautionary actions, stockpiling essential products to mitigate the risk of a possible stock shortage, which is often called "panic buying" [27]. The term is defined as the phenomenon that occurs when consumers buy unusually large amounts of products in anticipation of, during or after a disaster or perceived disaster, or in anticipation of a large price increase or shortage. It is a specific herd behaviour that is mainly triggered by a disaster or health crisis [26].

It has the potential to disrupt the supply chain with increased demand [27]. This disruption leads to more panic buying, thus creating a vicious circle. Moreover, panic buying reduces supply and creates higher demand, leading to higher price inflation. This increase in demand leads to a shortage of the product.

There are three mechanisms that can cause panic buying [28]. Firstly, it could be a manifestation of conflict between the desire to maintain routines versus uncertainty regarding the duration of the pandemic limiting access to daily necessities. Secondly, when a high risk is perceived, it is more likely for consumers to carry out panic buying in order to minimize their stress. In this regard, panic buying can be viewed as a self-protection mechanism to satisfy their safety needs [26]. Lastly, it could be a reaction in response to one's loss of control of the future, causing people to react by conducting social behaviours that are similar to that of other consumers.

The recent cases of panic buying were not a novelty. There are previous pandemics or natural disasters such as earthquakes or hurricanes, that have prompted waves of panic buying [29–32].

1.2. The Food Attributes and Situational Factors That Influence Panic Buying

Several approaches have been adopted to model the buying behaviour of food consumers. Of these, the multi-attribute approaches are based on the assumption that quality is a multidimensional phenomenon [33]. Intrinsic cues are those that are associated with the physical properties of the product such as taste and flavour, whereas extrinsic attributes are all others, such as brand name and the reputation of the seller [34].

When panic buying, the importance that consumers assign the different product attributes is modified [35]. Thus, quality properties are usually less important than the amount [2], and consumers tend to be more accepting of high prices for the products [36], as long as they can secure the supply.

The situational factors that often affect shopping are altered [2]. Thus, the information transmitted by "reliable" sources is considered "cheap talk" and sometimes contributes to increase panic, meaning that the most "credible" information can be people's experience when seeing empty shelves. Authors suggest that there can be two types of problems with this type of buying: coordination failure (all consumers buy at the same time) and information failure (consumers do not know the supply chain and they believe that the shortage can last a long time). Furthermore, in this digital era, information is readily available and can be quickly disseminated to masses over multiple channels that are also susceptible to abuse [26].

On the other hand, consumers' buying decisions are often influenced by the choices of their peers, and this is more obvious in panic buying [37] where substitute products can be more readily accepted [38].

Panic buying is also affected by the measures that, both to mitigate stockpiling as well as to decrease the risk of contagion, may have been implemented at the places of purchase [39]. Among the measures established to prevent stockpiling are limiting the items that a single person can buy, shorter opening hours and information campaigns to deter stockpiling. Furthermore, to prevent possible contagion in the establishment, the most noteworthy measures are the distance between people, separation with screens, the recommendation to go shopping alone, following signs on the ground that keep the person shopping from going backwards, paying with credit cards, or distributing cleaning and protection material among the customers [29].

One of the situational factors that most affects buying and which has been modified by the implemented measures, is the time that can be allocated to it. Thus, it has been verified that time pressure changes the factors that affect buying food [39]. Shoppers under time pressure are less likely to make unplanned purchases compared to those who are not under time pressure [40], and this factor is an important determinant of aspects such as reading labels [41]. Under time restrictions, consumers have more trust for high-priced products and high-quality brands [42].

Furthermore, new technologies have really boomed during this crisis, and as well as enabling access to different sources of information [43], they have also modified distribution channels. The online channel, which had hardly been used for food until now, has become popular [20]. Moreover, it has been a very useful tool that limits the physical contact which is common when shopping [44].

1.3. The Consumer and Panic Buying

The panic buying behaviour does not affect all consumers equally. In general, it increases if consumers have previously experienced similar issues, and it decreases as the stock builds up [45]. Having experienced these issues previously is not essential however, as the effect of observing what happens to others is decisive [46].

It has also been observed that people with high levels of anxiety can go panic buying more often and stockpile more products. However, people with low levels of anxiety can also be dangerous because they are less likely to conduct the necessary actions to contain the pandemic [47]. Fear motivates people to go shopping because it gives them a sense of security and it alleviates their stress. It is a way of keeping their negative emotions under control [48,49].

Panic buying happens after numerous personal decisions in a short amount of time, which makes them especially difficult to research. However, it is important to do so because they can have a very negative effect. Panic buying is troublesome, and its consequences mainly affect vulnerable groups of people who cannot access essential goods such as food or water [3,26,27].

Furthermore, they also affect companies in the sector and supply chains. The food sector faced an increased demand due to the panic-buying and stockpiling of food. Disruptions have been minimal thus far, as food supply has been adequate, and markets have been stable. However, we have already seen challenges in terms of logistic bottlenecks, and there is likely less food of high-value commodities (i.e., fruits and vegetables) being brought to the market [50]. The supply of some products has been affected, forcing food companies and distribution chains to make significant efforts.

2. Materials and Methods

2.1. Database

For this study, Spanish buyers were surveyed using social networks Facebook and WhatsApp between 2 and 14 April 2020. It is a non-discriminatory exponential snowball sampling. The characteristics of the study and the state of alarm in Spain made it difficult to apply other sampling methods. A single questionnaire was submitted to each consumer during the lockdown, posing questions linked to the pre-lockdown week as well as the first week of the lockdown. The different sections that were in the questionnaire are listed in the methodology section.

We collected 528 valid responses. A confidence interval of 95.5% ($k = 2$) was selected, as this value is the most common one used in socioeconomic studies. The sample error was calculated within an infinite population (total food consumers in Spain) and assuming maximum indetermination (as it is usually done, with $p = q = 0.5$).

Before the fieldwork, a preliminary questionnaire was sent to 10 consumers to confirm that the questions of the survey were well designed and easily understandable.

A questionnaire with five parts was designed. The first included general buying habits before COVID-19. The second included questions on the level of concern and information on the health crisis. In the third, the questions addressed buying food the week prior to the state of alarm (7 to 13 March), and in the fourth, buying food during the first week of lockdown (14 to 20 March). Lastly, Section 5 includes the socio-economic data of the people polled, such as their gender, age, level of education, net monthly family income, and location of residence. The questionnaire used is included in the Supplementary File.

Answers from almost all the autonomous communities of Spain were gathered, except for the Canary Islands and Asturias. However, the answers from the communities of Murcia, the Valencian Community, and Castile-La Mancha stand out because, by using snowball sampling, it has been affected by the location of origin of the researchers, who posted the questionnaire on their social networks. Figure 1 shows the distribution of the sample according to the community of origin.

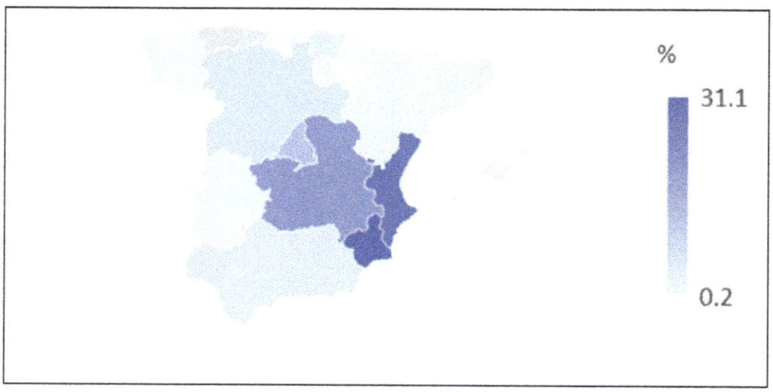

Figure 1. Sample distribution by autonomous community of origin.

In general, the people in charge of household shopping are mainly characterised by being women aged 35 to 49, with university studies, net monthly family incomes between €1000 and €1999 and who live in areas of high urban population concentration (Table 1).

Table 1. Socio-economic characteristics of the sample of people who are in charge of shopping.

Variables	%
Gender	
Male	32.0
Female	68.0
Age (in Years)	
18–24	5.1
25–34	20.6
35–49	44.5
50–64	26.5
>65	3.3
Education Level	
Elementary	2.1
Secondary	13.6
University	84.3
Income	
<€1000	6.4
€1000–€1999	30.9
€2000–€2999	28.0
€3000–€3999	20.3
>€4000	14.4
Area of Residence	
Rural (<30,000 inhab.)	19.8
Urban (30,000–100,000 inhab.)	16.2
High urban population concentration (100,001–500,000 inhab.)	53.4
Metropolitan area (>500,000 inhab.)	10.6

2.2. Statistical Analysis

A segmentation of the sample population was conducted to analyse the information. The variable selected for segmentation was the variation of the importance of purchase attributes in this pandemic situation. To create this new variable, the importance that every individual consumer gave to the perceived importance of different food purchasing attributes in the pre-lockdown week and during the lockdown were compared. The attributes included were price, origin, place of purchase, type and size of packaging, commercial brand, organic certification, and designation of origin. The people polled were required to rate the importance attributed to the reported purchase attributes on a 5-point Likert scale, with 1 being the least important and 5 the most important in their day-to-day shopping. With that information, a new variable was established for each attribute, with three possible scores depending on the amount it changed: 1, if it did not change (a 3 rating on the Likert scale); 2, if it changed slightly (rating of 2 and 4 on the Likert scale, slightly less and slightly more, respectively); and 3, if it changed significantly (rating of 1 and 5 on the Liker scale, a lot less and a lot more, respectively). In this regard, three segments were identified in accordance with the variation in the assessment of purchase attributes of consumers due to the pandemic. Segment 1: No change in purchase attributes; Segment 2: Slight change in purchase attributes; and Segment 3: Major change in purchase attributes.

Previously, other studies have used the purchase attribute assessment to determine consumer segments [51–53]. Only extrinsic attributes have been considered in this case, as the study asks about food in general, and not a single product. Extrinsic attributes are product-related attributes which are not a part of the physical product, meaning they can be changed without altering the properties of the physical product [34]. Selecting these attributes was done by taking into account those which generally have a greater influence when buying food and which are most susceptible to be modified by panic buying. The attached table shows a summary of the extrinsic attributes taken into account and their consideration in other research papers (Table 2).

Table 2. Extrinsic attributes considered in some food research studies.

Attribute	References
Price	[54–61]
Origin	[57,58,62–68]
Place of purchase	[69–76]
Type and size of packaging	[77–85]
Commercial brand	[55,59,86–93]
Organic certification	[54,55,64,86,92,94–98]
Designation of origin (guarantee label)	[63,76,87,99–107]

In order to look for differences among segments, we created a cross-tabulation table with a column proportions test for nominal variables and an Anova test to analyse scale and ordinal variables. The statistical tool used to contrast the hypothesis of independence between categorical variables was Pearson's chi-squared test. Specifically, the cross-tabulation tables and chi-squared test were used to analyse the existence of significant differences between segments regarding different variables.

Results were analysed using the Statistical Package for Social Sciences IBM SPSS version 25.

3. Results and Discussion

The pandemic changed the importance that some consumers gave to different food purchasing attributes. Our segmentation identifies three different groups of consumers attending to the importance of those changes. The first segment, with 38.6% of the sample, included consumers who had not modified the importance assigned to the food purchasing attributes before and during the lockdown. The second segment, with 47.5% of the sample was comprised by consumers who showed slightly modifications. Finally, the third segment of consumers, with 13.9% of the sample, included those who showed severe changes.

3.1. Amount Purchased of Different Food Items

The variation of the amounts of different food items purchased by the consumer segments the week prior and after the state of alarm are shown on Table 3. This table shows the average scores for the amounts purchased before the state of alarm compared to those purchased in a normal situation. Thus, a one means that they buy much less than usual and a five means that they buy a lot more than usual. Therefore, scores over three indicate that more has been bought, and lower scores indicate that less has been bought.

In general, the food items purchased the most in the week prior to the state of alarm were rice, pasta and legumes, dairy products, meat, fresh fruit and/or vegetables, baked goods, and frozen foods. These are basic necessity foods and many of them are easily storable. Of these, in week 1 of the lockdown, the increase in dairy products, meat, fruit and vegetable and baked goods remained the same and the purchase of rice, pasta and legumes, and frozen products fell. The least purchased food items before the lockdown were soft drinks and juices, spices, condiments and sauces, snacks, bottled water, and beer, wine, and spirits. Of these, the purchase of beer and snacks increased in week 1. The purchase of bottled water and soft drinks remains the same. Lastly, the foods that remained the same before the lockdown are coffee and infusions, olive oil, fish, and canned products. All of the latter fell in week 1 of the lockdown. Regarding olive oil, an essential part of the Mediterranean diet, it was not added in a significant way when panic buying before the state of alarm, as all segments have ratings of around 3.

Table 3. Amount of different food groups purchased in the pre-state of alarm week and in which direction the change went in the first week post-state of alarm for each of the consumer segments.

Food	Seg. 1 No Change (38.6%) [1]			Seg. 2 Slight Change (47.5%) [1]			Seg. 3 Major Change (13.9%) [1]			Total		
Rice, pasta, legumes ***	3.30	***	↓	3.52	**	↓	3.56	***	↓	3.44	***	↓
Dairy products **	3.18			3.36	**	↑	3.37			3.29	**	↑
Meat **	3.14	*	↑	3.28			3.38			3.24	*	↑
Fresh fruit and vegetables ***	3.10			3.27	***	↑	3.36			3.22	***	↑
Baked goods *	3.12			3.21	***	↑	3.29			3.18	***	↑
Frozen foods	3.08			3.11	*	↓	3.04			3.09	***	↓
Coffee and infusions	2.98			3.06			3.04	**	↓	3.03	***	↓
Olive oil	3.00	*	↓	3.02			2.96	*	↓	3.00	***	↓
Fish	2.99	**	↓	2.94			2.90			2.95	**	↓
Canned food	2.93	**	↓	2.98	***	↓	2.93	**	↓	2.95	***	↓
Beer, wine, and spirits ***	2.94	*	↑	2.93			2.59			2.89	**	↑
Bottled water	2.92			2.86			2.84			2.88		
Snacks *	2.90	*	↑	2.92	*	↑	2.64			2.88	**	↑
Spices, condiments, and sauces *	2.77			2.76			2.55	**	↓	2.74	**	↓
Soft drinks and juices ***	2.71			2.76			2.33			2.68		

[1] Size of the segment; ***, ** and * indicate significant differences with a 1%, 5%, and 10% margin of error; the asterisks shown on the food column reflect differences among segments. The asterisks shown in the segment column refer to the differences between the week prior and the first week after the state of alarm for each segment; the arrows mark the direction of the variation ↓ the amounts purchased decreased in week 2 compared to week 1. ↑ The amounts purchased increased in week 2 compared to week 1.

In segment 1, which comprised 38.6% of consumers, the most purchased foods in the pre-lockdown week were the same as for the total population, although the amount purchased was slightly lower in all cases: dairy products, baked goods, meat, fruit, rice, pasta, and legumes. Of these, in week 1 of the lockdown, only the purchase of meat increased, whereas rice, pasta and legumes decreased, with this fall being more significant than for other segments. Of the least bought products, snacks and beer, their purchase increased the following week, and it did so in greater proportion than for the total population. For the foods that remained the same, the amount purchased decreased in week 1 of the lockdown for oil, fish, and canned food, whereas for coffee it remained similar.

In segment 2, which comprised 47.5% of consumers, the food items that were purchased the most were similar to the total population, and the amounts purchased were slightly higher than the total population. The variation in week 1 of the lockdown was also similar. Regarding the products whose amount purchased dropped, they were also similar, and in week 1 the only group that increased were snacks, with the amount purchased for all other products remaining the same. For the food items that remain more stable, the only group that was purchased less was canned food, with all other categories remaining the same. This is the intermediate segment, which reacted by modifying the amounts purchased, but in a more moderate way than segments 1 and 3.

Lastly, segment 3, which comprised 13.9% of consumers, is the most sensitive to change. The increase in the amount purchased the week prior to the lockdown took place for the same groups of products as for the total population, but in this case, the increase was more pronounced. In week 1 of the lockdown, the amounts purchased were similar for all groups of products, except for rice, pasta, and legumes. The products purchased at a lesser rate were similar to the rest of the population, but it is worth noting that the purchase of beer, wine, and snacks was much lower during the week prior, and these levels were maintained in week 1 of the lockdown. This segment also purchased less spices, condiments, and sauces, and during week 1 they purchased even less of these. In the categories of more stable products, there was a decrease in the amounts of coffee, oil, and canned food purchased. The amount of fish purchased remained lower than normal but stable during that period of time.

3.2. Food Buying and Places of Purchase

As regards shopping, a majority of consumers leave the home to do it, but this proportion was noticeably lower in segment 3, where the online or telephone option was much more prominent than for the other segments. When asked if they conducted these ways of shopping before COVID-19, the answers were very similar in all segments, and the percentage of those who did it often or on a regular basis was very low. However, this crisis boosted these ways of shopping, especially among consumers in segment 3. However, when asked whether they will continue using online shopping, in this segment there was a smaller percentage of consumers that will do so compared to others, where almost 30% of consumers said they will use the online channel to buy food. These results can be found in Table 4.

Table 4. Traditional shopping versus online food shopping for home.

Food Shopping for Home	Seg. 1 (38.6%) [1]	Seg. 2 (47.5%) [1]	Seg. 3 (13.9%) [1]	Total
Buying Food ***				
Yes, I leave home to shop	89.2% ^{a}	87.3% ^{a}	72.6% ^{b}	86.0%
No, I make a list and a friend/relative brings it to me	7.8% ^{a}	7.6% ^{a}	9.6% ^{a}	8.0%
No, I do it over the Internet	2.5% ^{a}	3.2% ^{a}	12.3% ^{b}	4.2%
No, I do it over the telephone	0.5% ^{a}	2.0% ^{a,b}	5.5% ^{b}	1.9%
Online Food Purchasing (Before the State of Alarm)				
Never	63.7% ^{a}	61.0% ^{a}	72.6% ^{a}	63.6%
Rarely	18.6% ^{a}	19.1% ^{a}	8.2% ^{b}	17.4%
Sometimes	15.2% ^{a}	17.5% ^{a}	17.8% ^{a}	16.7%
Most of the time	2.0% ^{a}	1.6% ^{a}	0.0%	1.5%
Always	0.5% ^{a}	0.8% ^{a}	1.4% ^{a}	0.8%
Online Food Purchasing (After the State of Alarm)	30.40% ^{a}	35.10% ^{a}	47.90% ^{b}	35.00%
Consumers that Will Continue Buying Online				
Definitely not	33.9%	27.3%	45.7%	33.0%
I don't know	37.1%	45.5%	40.0%	41.6%
Definitely will	29.0%	27.3%	14.3%	25.4%

[1] Size of the segment; *** note significant differences with a 1% margin of error, respectively; Different letters in the same row mean significant differences for the segments ($p < 0.05$).

Regarding the places of purchase (Table 5), the scale used was an ordinal scale of five items, where 1 means that the person never purchased at that type of establishment, and a five means that they always did. The establishment of choice was the supermarket or hypermarket, followed by traditional shops, indoor markets and, lastly, buying straight from the producer, and there were hardly any differences among segments. Regarding the differences in behaviour between the week prior to the lockdown and during the lockdown, there were differences in almost all cases. The supermarket/hypermarket as the preferred choice increased mainly for segment 1, whereas it decreased for segment 3, although not significantly. All other places of purchase suffered a drop in importance.

Table 5. Place of purchase of food for home.

Place of Purchase	Seg. 1 (38.6%) [1]	Seg. 2 (47.5%) [1]	Seg. 3 (13.9%) [1]	Total
Supermarket/Hypermarket				
Pre-state of alarm	4.06 *	4.05	4.21	4.08
Post-state of alarm	4.24 *	4.06	4.03	4.12
In Traditional Shops				
Pre-state of alarm	2.94 *	3.13 *	3.14 *	3.06
Post-state of alarm	2.61 *	2.79 *	2.71 *	2.71
In Indoor Markets/Markets				
Pre-state of alarm ***	2.15 *	2.36 *	2.10 *	2.24
Post-state of alarm	1.50 *	1.73 *	1.62 *	1.63
Directly from the Producer				
Pre-state of alarm	1.39 *	1.38 *	1.34	1.38
Post-state of alarm	1.16 *	1.21 *	1.27	1.20

[1] Size of the segment; *** and * note significant differences with a 1% and 10% margin of error, respectively; the asterisks in the segment boxes refer to the change between considered periods (*t*-test of related samples). The asterisks on the variables column refer to the differences between segments (ANOVA).

3.3. Purchase Attributes

The most highly valued attribute was the origin, followed by the price, place of purchase, size of the packaging, the protected designation of origin (PDO) label, the type of packaging, and the organic certification of the food. The brand proved to be less important. The attributes that show significant differences between the consumer segments were the price, brand, and organic nature of the product. Consumers in segment 1 valued the price less and the organic nature slightly less, whereas, regarding the brand, they were in line with the rest of the population. Segment 2 consumers valued the organic nature and brand more, whereas those in segment 3 stood out for assigning greater value to the price, and less to the brand (Figure 2).

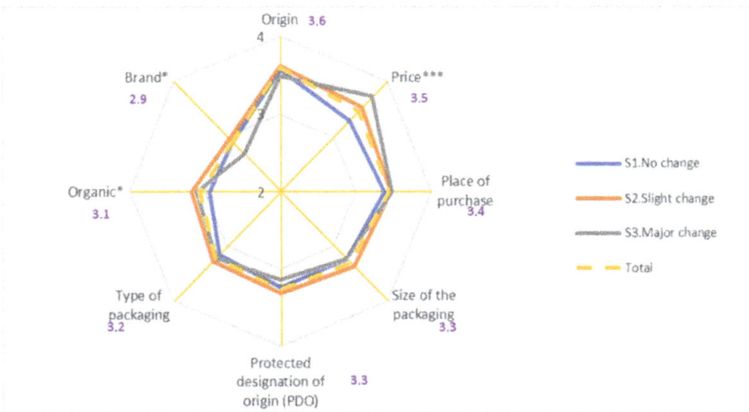

*** and * significant differences with a 1% and 10% margin of error.

Figure 2. Attribute value by consumer segments.

Regarding the direction of the change in importance assigned to the attributes, segment 1 has not been considered, as it did not modify the value it assigns to the attributes (Figure 3). In general, in segment 3 there was a greater decrease in the importance assigned to purchase attributes.

If we analyze them individually, the brand is the attribute that most decreased in importance, especially for segment 3. However, regarding the designation of origin, although its decrease was also majoritarian, there was a 24.7% of consumers in segment 3 who assigned it greater importance after the lockdown. This also happened with organic certification.

The place of purchase was the attribute with the most significant increase. In segment 2, there was a high percentage of people who maintained its importance, and in segment 3, compared to the other attributes, its importance decreased for a smaller percentage of consumers.

The size of packaging is the second attribute for which a greater proportion of consumers increase its importance. Regarding the origin, there was a large percentage of people in segment 3 who said its importance decreases. However, in segment 2, the percentage of people who assigned it the same level of importance was very high.

The price is the attribute that most decreased in importance for segment 2, whereas for segment 3, there was a greater percentage of people who maintained its importance compared to the other attributes.

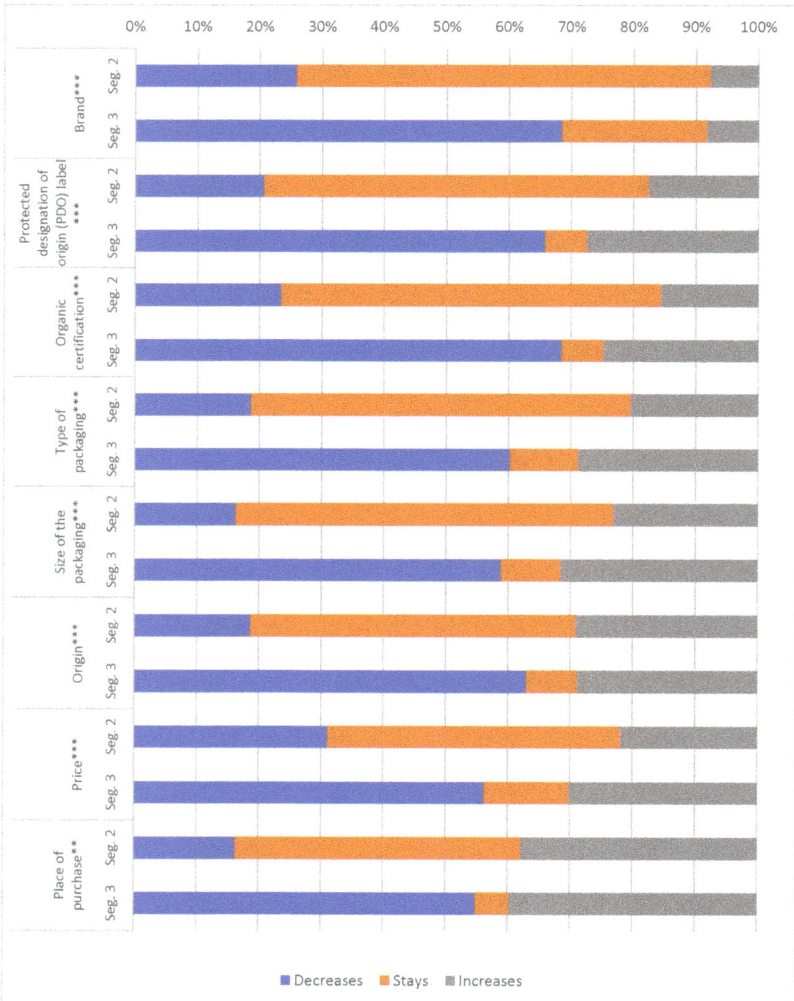

*** and ** notes significant differences with a 1% and a 5% margin of error.

Figure 3. Change in the value assigned to the attributes.

3.4. Level of Concern for COVID, Impact, and the Search for Information

The level of concern was very high and there were no significant differences in this variable (Table 6). Regarding the impact, it was fairly negative for all segments, especially for segment 3. Consumers in segment 1 were more optimistic.

Table 6. Level of concern and impact of the COVID-19 crisis in Spain.

Level of Concern and Impact of the COVID-19 Pandemic	Seg. 1 (38.6%) [1]	Seg. 2 (47.5%) [1]	Seg. 3 (13.9%) [1]	Total
Level of Concern				
Not at all concerned	1.0%	0.0%	1.4%	0.6%
Slightly concerned	2.5%	0.8%	1.4%	1.5%
Somewhat concerned	14.2%	9.6%	8.2%	11.2%
Moderately concerned	32.4%	36.7%	28.8%	33.9%
Extremely concerned	50.0%	53.0%	60.3%	52.8%
Impact on the Family Economy ***				
Very negative	13.7% [a]	24.3% [b]	35.6% [b]	21.8%
Somewhat negative	55.4% [a]	47.0% [b]	37.0% [b]	48.9%
Neutral	27.9% [a]	23.1% [a]	20.5% [a]	24.6%
Somewhat positive	2.5% [a]	4.8% [a]	4.1% [a]	3.8%
Very positive	0.5% [a]	0.8% [a]	2.7% [a]	0.9%

[1] Size of the segment; *** notes significant differences with a 1% margin of error; Different letters in the same row mean significant differences for the segments ($p < 0.05$).

The results reveal a high concern for the pandemic, and only 1.5% of consumers were not interested in the information. Consumers in segment 3 were more concerned and also spend more time seeking information, especially through the news programs on TV and official sources. The opinion of friends and family and social networks are less important, although they are taken into account by segment 2 (Table 7),

Table 7. Time spent seeking information on COVID-19 (h/day) and preferred sources of information.

Time and Sources of Information	Seg. 1 (38.6%) [1]	Seg. 2 (47.5%) [1]	Seg. 3 (13.9%) [1]	Total
Time Spent Seeking Information (hrs/day)				
I'm not interested in being informed	2.5% [a]	1.2% [a]	0.0%	1.5%
Less than 1 h	37.3% [a]	35.9% [a]	35.6% [a]	36.4%
Between 1 and 2 h	44.6% [a]	40.6% [a]	34.2% [a]	41.3%
Between 2 and 4 h	9.8% [a]	15.5% [a,b]	21.9% [b]	14.2%
More than 4 h	5.9% [a]	6.8% [a]	8.2% [a]	6.6%
Sources of Information				
TV news ***	3.49	3.70	3.36	3.57
Newspapers (in paper or online)	3.29	3.49	3.45	3.41
Relatives and/or friends	3.11	3.22	2.99	3.15
Official sources ***	2.95	3.18	3.29	3.10
Social networks	2.97	3.14	2.93	3.05

[1] Size of the segment; *** notes significant differences with a 1% margin of error; Different letters in the same row mean significant differences for the segments ($p < 0.05$).

3.5. Socio-Economic Characteristics of People Polled by Segments

Only the socio-economic characteristics with significant differences among segments are shown (Table 8). Only differences in age and family size emerged. Regarding age, in segment 3 there was a larger of percentage of younger consumers (aged 18 to 34). The segment of people whose behaviour did not change has a smaller family size. Regarding gender and work activity there were no significant differences.

Table 8. Age and family size of the consumer segments by their assessment of the purchase attributes.

Age and Family Size	Seg. 1 (38.6%) [1]	Seg. 2 (47.5%) [1]	Seg. 3 (13.9%) [1]	Total
	Age (in Years) **			
18–24	5.4% [a,b]	3.6% [b]	9.6% [a]	5.1%
25–34	19.6% [a]	17.9% [a]	32.9% [b]	20.6%
35–49	44.6% [a]	48.6% [a]	30.1% [b]	44.5%
50–64	25.5% [a]	27.9% [a]	24.7% [a]	26.5%
>65	4.9% [a]	2.0% [a]	2.7% [a]	3.2%
Family Size (members) *	2.84	3.12	3.01	2.99

[1] Size of the segment; ** and * note significant differences with a 5% and 10% margin of error, respectively; Different letters in the same row mean significant differences for the segments ($p < 0.05$).

4. Discussion

The results of the survey have made it possible to learn how food consumers behaved right before the lockdown and in the first week of the lockdown in Spain, and which consumer profiles changed their behaviour the most. This section analyses the results taking other research into account.

In our study, 61.4% of consumers modified their buying behaviour. Other studies, such as one conducted in China after the COVID-19 outbreak, showed that the pandemic had a psychological impact on 54% of consumers [108]. People with higher levels of anxiety can conduct panic buying more often and stockpile more products. This can have a harmful effect on the community, which may need these resources for other purposes. On the other hand, people who modify their behaviour less can also be dangerous because they are less likely to conduct the necessary actions to contain the pandemic [47].

During the period of time analysed, the amounts purchased have changed. This can be due to the fact that consumers react when they believe that products will be scarce in order not to have their ability to choose limited [26]. Furthermore, other authors say that an increase in shopping can decrease stress before an unknown situation [48,49].

Stockpiling products the week before the lockdown also occurred in other countries, but it was slightly different. In Italy, the storable products that were in plastic packaging, which are perceived as being safer, were also bought more often [19]. However, the purchase of fresh products decreased, which did not happen in Spain. Concern for a healthy diet may be behind this behaviour [43]. However, the purchase of rice, pasta, and legumes and frozen products fell, possibly due to the fact that they were already stored in the homes. On the other hand, there were products that were purchased less in the week before the lockdown, but which were purchased more during the lockdown, such as beer or snacks. These are products that were usually consumed in bars or restaurants and are now consumed at home, and it may be because their consumption is linked to their symbolic value and the tendency to continue some external socialisation habits at home [19]. In our segments, segment 3 (panic buyers) is the one that buys the least snacks, beer, and wine, meaning it has not transferred social activities to the household or has not made prize product purchases. On the contrary, segment 1 modified its behaviour the least, except for beer and snacks, meaning it could be the segment that, while experiencing the least amount of alarm, needed to take its socialisation habits home the most. On the other hand, beer and snacks are products that can be considered "prize products", whose purchase increases as a result of a disaster [32]. Further noteworthy, is the drop in the purchase of olive oil in the first week of the lockdown, which could be due to it being a product with a high caloric value and consumers trying to decrease its intake [43].

Regarding the places of purchase, online purchasing, which until now had been a seldom used channel by food buyers, shows a significant increase at the expense of the conventional channels. This result is in line with those obtained in studies conducted on previous crises [109–111] as well as more recent ones focused on the COVID crisis [19,20]. The loss of importance of all other purchasing establishments has also been documented in other countries such as Italy [19].

In general, food attributes have fallen in importance, especially for panic buyers, which can be in line with the studies that say that, when panic buying, substitute products are more acceptable, in other words, consumers are more content with the products they find and do not carry out "specific searches" [38]. When analysing the evolution of these attributes individually, it is observed that the place of purchase of the attribute that has increased its importance for a greater proportion of consumers. This can be linked to other attributes derived from shopping. In general, consumers seek less contact, which is why they try to go shopping less often and seek safe places to do so [111]. During the two periods of time analysed, there is a loss of importance of the brand as a purchase attribute. Even though this contradicts one study [42], can be in line with another [38], which reveals the better acceptance of substitutes in panic situations. On the other hand, both the designation of origin and the organic label increased in importance during the lockdown for a significant percentage of consumers. It could be that established brands, which are well positioned regarding the perception of quality of their products, work better than private brands, and that the result is in line with some authors [29], which says that well-positioned brands will have an advantage during pandemics. Furthermore, authors [112] says that in situations of risk, consumers prefer to buy organic products. The price, which decreased in importance for segment 2 and remained stable for segment 3, in other studies [42] was seen as an indicator of the quality of the product, meaning it is a significant attribute, while revealing that a high price increases consumer confidence. Furthermore, when a product shortage is predicted, the consumer tends to be more accepting of higher prices for products [36]. The changes in the importance assigned to the size of the packaging can also be linked to the lower frequency with which people go shopping [111]. On the other hand, if we take into account that people often go shopping in order to store the food, both the type of packaging and its size can be very important. Lastly, the decrease in importance of the origin attribute for panic buyers and its preservation among consumers in segment 2 are in line with the findings of other studies [20,111].

Regarding the level of concern for Covid, a study [112] believes that those consumers who perceive greater risks change their buying behaviours more, which has also been observed in our results (Table 7). However, the fact that consumers in segment 3, who perceived a greater risk, still consider the price important, can be in line with the consideration that price restrictions are a consequence of the foreseeable economic crisis [29].

Regarding the sources of information, media outlets play a very important role to provide credible information to consumers on how the supply chain works, and thus alleviate the issues derived from stockpiling [2] as well as favouring an increase in confidence among consumers in the public authorities, while possibly encouraging a return to the normal buying behaviour [46]. On the other hand, social networks and the opinion of friends and family, despite being less important in this study, is direct information that can have a great influence on behaviour [46]. How prepared people are for disasters should take into account the disastrous events that are likely to happen and what/who is likely to be affected in different parts of countries and cities [12]. This can be done with suitable information from official channels, but also using social networks, which are gaining traction as media outlets among consumers.

The results obtained regarding socioeconomical variables show that in the segment of panic buyers there is a greater proportion of young people, which is consistent with the fact that the COVID crisis has an effect regarding changing behaviour that is less significant among older people [112]. In segment 1, which includes those who experience less changes in their behaviour, the size of the family is smaller, which is in line with the concern of the consumers with their family becoming infected [108]. Lastly, in our study there were no differences regarding gender and work activity, whereas in another study there were differences in these variables, having a greater impact on women and students [108].

5. Conclusions

The work conducted has made it possible to detect three significant segments of consumers based on their food buying behaviour, which also made it possible to establish the variations experienced both regarding the place of purchase as well as the extrinsic attributes of the food items, as well as the profile of the consumers in both analysed periods.

Up to 61.4% of consumers modified their buying behaviour, with the consumers that most modified the value they assign to the attributes being those who stockpiled the most (panic buying), whereas those who changed it the least, shopped as usual (38.6%).

The most valued attributes were the origin, followed by price, place of purchase, size of packaging, protected designation of origin label, type of packaging, and organic certification of the food, in this order. It is worth noting the low importance that consumers assigned to the brand, maybe due to the fact that what was important was having supplies.

Regarding the attributes with significant differences among consumer segments, we see that whereas consumers in segment 1 (38.6% of consumers) assigned less value to the price and organic certification, and more to the brand, consumers in segment 2 (47.5% of consumers) assigned greater value to the organic certification and the brand. Lastly, consumers in segment 3 (13.9% of consumers) assigned greater value to the price and less to the brand.

In this sense, the study has verified that the importance assigned to purchase attributes has been modified to a greater or lesser extent, especially for the brand, which can indicate that the consumer has more readily accepted substitute products. However, attributes such as the protected designation of origin or organic certification, which are linked to quality assurance and food safety increased their importance, especially among consumers who were more prone to change, but not the price, which was different for each segment, gaining importance among consumers who were more prone to panic buying, possibly because they also perceived greater risk from the health, economic and social crisis.

Further noteworthy, is the change in place of purchase of the food, as it is observed that supermarkets and hypermarkets, where it is more feasible to find a broader range and in greater amounts for storing, benefited. In this sense, online food buying, which until before the pandemic had very limited importance, has now gained traction and many consumers expressed their willingness to continue using this way of shopping.

Lastly, it has been verified that the official channels of information and written press (in paper and online) are the most reliable sources of information that reach consumers and also modify their buying behaviour the most, whereas over social networks it is possible to reach consumers who have maintained a more stable buying behaviour.

The characterization of the profile of the different consumer segments can allow food production and/or distribution companies to implement different innovative, customized, and more effective strategies, aimed at decreasing the impact of stockpiling, and therefore of food shortage. To do so, the alternative strategies they can implement include a stock of non-perishable foods as a strategic reservoir in their warehouses in preparation for eventual increases in demand, to increase production capabilities in a sustainable way, and especially, to promote the online sale and distribution of food, with the objectives of lowering the amount of people in shops (which decreases the spreading of the pandemic and favours health) and preventing consumers from encountering possible circumstantial shortages that would only encourage stockpiling and panic buying, even among consumers who do not change their buying behaviour.

The goal was for the results of the research to discover the changes in buying behaviour of food consumers in situations of severe stress, due to the COVID-19 pandemic, compared to their regular behaviour. In turn, this situation has led to an economic and social crisis that only the income policy of the government has attempted to alleviate, in order to prevent it from becoming a financial and credit crisis.

Unfortunately, this crisis has shown countries, such as Spain and others in Europe that do not suffer major natural disasters like other countries in the world that routinely witness hurricanes, earthquakes, major fires, floods, droughts, epidemics, etc., that they must be prepared, as they are increasingly frequent in other parts of the world where they did not use to happen, having a significant impact on agricultural production, which is the basis for the production of foods.

Without a doubt, the increased frequency of natural disasters is caused by climate change, as has been revealed by the Intergovernmental Panel on Climate Change (IPCC). Furthermore, the transportation of people around different parts of the world has contributed to disseminate what were once local epidemics, making them global pandemics, as has happened this year with COVID-19, and the possibility that other pandemics may occur in the future cannot be ruled out..

Having said this, the goal of this study is not just to be prepared for future pandemics, but also to be vigilant for future situations that may occur and cause stress among consumers, panic buying, food stockpiling and, potentially, shortages. Being prepared is the responsibility of the private sector (from the standpoint of logistics), as well as the public authorities (from the standpoint of truthful information for the population). The ultimate goal is to care for the basic needs of the population, one of which is eating.

Only when ready is it possible to respond appropriately. What is very probable is that many of these changes in the buying behaviour of consumers will happen again in situations of health, economic, financial and/or social crises. Knowing and identifying when they take place will make it possible to plan for them in advance, making quick and useful decisions from the field of the agri-food sector.

Supplementary Materials: The following are available online at http://www.mdpi.com/2304-8158/9/12/1821/s1, Supplementary File: Food buying habits during a crisis: Covid-19.

Author Contributions: Conceptualization, M.B., L.M.-C., A.R. and R.B.; Formal analysis, M.B., L.M.-C., A.R., R.B.; Methodology, M.B. and L.M.-C.; Resources, M.B. and L.M.-C.; Software, M.B.; Supervision, M.B., L.M.-C., A.R. and R.B.; Writing—original draft, M.B., L.M.-C., A.R. and R.B.; Writing—Review & Editing M.B., L.M.-C., A.R. Visualization: M.B. All authors contributed equally to this manuscript. All authors have read and agreed to the published version of the manuscript.

Funding: This research received no external funding.

Conflicts of Interest: The authors declare no conflict of interest.

References

1. Taleb, N.N. *The Black Swan: The Impact of the Highly Improbable*; Random House Publishing Group: New York, NY, USA, 2007.
2. Mohan, V.; Berg, C.; Poblet, M. Panic, Information and Quantity Assurance in a Pandemic. Available online: https://ssrncom/abstract=3575232 (accessed on 14 April 2020).
3. Nicola, M.; Alsafi, Z.; Sohrabi, C.; Kerwan, A.; Al-Jabir, A.; Iosifidis, C.; Agha, M.; Agha, R. The socio-economic implications of the coronavirus pandemic (COVID-19): A review. *Int. Surg. J.* **2020**, *78*, 185. [CrossRef] [PubMed]
4. Morse, S.S.; Mazet, J.A.; Woolhouse, M.; Parrish, C.R.; Carroll, D.; Karesh, W.B.; Zambrana-Torrelio, C.; Lipkin, W.I.; Daszak, P. Prediction and prevention of the next pandemic zoonosis. *Lancet* **2012**, *380*, 1956–1965. [CrossRef]
5. Woolhouse, M.; Scott, F.; Hudson, Z.; Howey, R.; Chase-Topping, M. Human viruses: Discovery and emergence. *Philos. Trans. R. Soc. B* **2012**, *367*, 2864–2871. [CrossRef] [PubMed]
6. Inayatullah, S. Neither A Black Swan Nor A Zombie Apocalypse: The Futures of a world with the Covid-19 coronavirus. *J. Future Stud.* **2020**, 18. Available online: https://jfsdigital.org/2020/03/18/neither-a-black-swan-nor-a-zombie-apocalypse-the-futures-of-a-world-with-the-covid-19-coronavirus/ (accessed on 15 September 2020).
7. Raoult, D.; Mouffok, N.; Bitam, I.; Piarroux, R.; Drancourt, M. Plague: History and contemporary analysis. *J. Infect.* **2013**, *66*, 18–26. [CrossRef]

8. Horrox, R. *The Black Death*; Manchester University Press: Manchester, UK, 2013.
9. Geddes, A.M. The history of smallpox. *Clin. Dermatol.* **2006**, *24*, 152–157. [CrossRef]
10. Taubenberger, J.K.; Reid, A.H.; Fanning, T.G. The 1918 influenza virus: A killer comes into view. *Virol. J.* **2000**, *274*, 241–245. [CrossRef]
11. WHO. Global Health Observatory Data. 2019. Available online: https://wwwwhoint/data/gho/data/indicators/indicator-details/GHO/estimated-number-of-people-(all-ages)-living-with-hiv (accessed on 15 September 2020).
12. Wai Man Fung, O.; Yuen Loke, A. Disaster preparedness of families with young children in Hong Kong. *Scand. J. Public Health* **2010**, *38*, 880–888. [CrossRef]
13. Sohrabi, C.; Alsafi, Z.; O'Neill, N.; Khan, M.; Kerwan, A.; Al-Jabir, A.; Iosifidis, C.; Agha, R. World Health Organization declares global emergency: A review of the 2019 novel coronavirus (COVID-19). *Int. Surg. J.* **2020**, *76*, 71–76. [CrossRef]
14. Anderson, R.M.; Heesterbeek, H.; Klinkenberg, D.; Hollingsworth, T.D. How will country-based mitigation measures influence the course of the COVID-19 epidemic? *Lancet* **2020**, *395*, 931–934. [CrossRef]
15. Thunstrom, L.; Newbold, S.; Finnoff, D.; Ashworth, M.; Shogren, J.F. The benefits and costs of flattening the curve for COVID-19 Forthcoming. *J. Benefit-Cost Anal.* **2020**. [CrossRef]
16. Kraemer, M.U.G.; Yang, C.-H.; Gutierrez, B.; Wu, C.-H.; Klein, B.; Pigott, D.M.; du Plessis, L.; Faria, N.R.; Li, R.; Hanage, W.P. The effect of human mobility and control measures on the COVID-19 epidemic in China. *Science* **2020**, *368*, 493–497. [CrossRef] [PubMed]
17. Lippi, G.; Henry, B.M.; Bovo, C.; Sanchis-Gomar, F. Health risks and potential remedies during prolonged lockdowns for coronavirus disease 2019 (COVID-19). *Diagnosis* **2020**, *7*, 85–90. [CrossRef] [PubMed]
18. Coetzee, B.J.S.; Kagee, A. Structural barriers to adhering to health behaviors in the context of the COVID-19 crisis: Considerations for low-and middle-income countries. *Glob. Public Health* **2020**, *15*, 1093–1102. [CrossRef]
19. Bracale, R.; Vaccaro, C.M. Changes in food choice following restrictive measures due to Covid-19. *Nutr. Metab. Cardiovasc. Dis.* **2020**, *30*, 1423–1426. [CrossRef] [PubMed]
20. Hall, M.C.; Prayag, G.; Fieger, P.; Dyason, D. Beyond panic buying: Consumption displacement and COVID-19. *J. Serv. Manag.* **2020**. [CrossRef]
21. Jribi, S.; Ben Ismail, H.; Doggui, D.; Debbabi, H. COVID-19 virus outbreak lockdown: What impacts on household food wastage? *Environ. Dev. Sustain.* **2020**, *22*, 3939–3955. [CrossRef]
22. Martínez-Carrasco, L.; Brugarolas, M.; del Campo, F.J.; Martínez-Poveda, Á. Influence of purchase place and consumption frequency over quality wine preferences. *Food Qual. Prefer.* **2006**, *17*, 315–327. [CrossRef]
23. Ajzen, I.; Fishbein, M. *Theory of Reasoned Action-Theory of Planned Behavior*; University of South Florida: Tampa, FL, USA, 1988.
24. Ajzen, I. *From Intentions to Actions Attitudes, Personality and Behavior*; Dorsey Press: Chicago, IL, USA, 1988.
25. Wierenga, B. Model and measurement methodology for the analysis of consumer choice of food products. *J. Food Qual.* **1983**, *6*, 119–137. [CrossRef]
26. Yuen, K.F.; Wang, X.; Ma, F.; Li, K.X. The Psychological Causes of Panic Buying Following a Health Crisis. *Int. J. Environ. Res. Public Health* **2020**, *17*, 3513. [CrossRef]
27. Dulam, R.; Furuta, K.; Kanno, T. Development of an agent-based model for the analysis of the effect of consumer panic buying on supply chain disruption due to a disaster. *Adv. Model. Simul. Eng. Sci.* **2020**, *7*, 102–116. [CrossRef]
28. Sim, K.; Chua, H.C.; Vieta, E.; Fernandez, G. The anatomy of panic buying related to the current COVID-19 pandemic. *Psychiatry Res.* **2020**, *288*, 113015. [CrossRef] [PubMed]
29. Martin-Neuninger, R.; Ruby, M.B. What Does Food Retail Research Tell Us About the Implications of Coronavirus (COVID-19) for Grocery Purchasing Habits? *Front. Psychol.* **2020**, *11*, 1448. [CrossRef] [PubMed]
30. Kurihara, S.; Maruyama, A.; Luloff, A. Analysis of Consumer Behavior in the Tokyo Metropolitan Area after the Great East Japan Earthquake. *J. Food Syst. Res.* **2012**, *18*, 415–426. [CrossRef]
31. Gómez, O.A. The Emergence of Food Panic: Evidence from the Great East Japan Earthquake. *J. Disaster Res.* **2013**, *8*, 814–825. [CrossRef]
32. Forbes, S.L. Post-disaster consumption: Analysis from the 2011 Christchurch earthquake. *Int. Rev. Retail Distrib. Consum. Res.* **2017**, *27*, 28–42. [CrossRef]

33. Grunert, K.G. What's in a steak? A cross-cultural study on the quality perception of beef? *Food Qual. Prefer.* **1997**, *8*, 157–174. [CrossRef]
34. Olson, J.; Jacoby, J. Cue utilization in the quality perception process. In Proceedings of the Third Annual Conference of the Association for Consumer Research, Chicago, IL, USA, 3–5 November 1972; Venkatesan, M., Ed.; Association for Consumer Research: Chicago, IL, USA, 1972; pp. 167–179.
35. Vancic, A.; Pärson, G.F.A. Changed Buying Behavior in the COVID-19 Pandemic: The Influence of Price Sensitivity and Perceived Quality. 2020. Available online: https://wwwdiva-portalorg/smash/get/diva2:1453326/FULLTEXT01pdf (accessed on 15 July 2020).
36. Yangui, W.; Hajtaïeb El Aoud, N. Consumer behavior and the anticipation of a total stockout for a food product: Proposing and validating a theoretical model. *Int. Rev. Retail Distrib.* **2015**, *25*, 181–203. [CrossRef]
37. Zheng, R.; Shou, B.; Yang, J. Supply disruption management under consumer panic buying and social learning effects. *Omega* **2020**, in press. [CrossRef]
38. Tsao, Y.C.; Raj, P.V.R.P.; Yu, V. Product substitution in different weights and brands considering customer segmentation and panic buying behavior. *Ind. Mark. Manag.* **2019**, *77*, 209–220. [CrossRef]
39. Suri, R.; Monroe, K.B. The effects of time constraints on consumers' judgments of prices and products. *J. Consum. Res.* **2003**, *30*, 92–104. [CrossRef]
40. Iyer, E.S. Unplanned purchasing: Knowledge of shopping environment and time pressure. *J. Retail.* **1989**, *65*, 40.
41. Park, C.W.; Iyer, E.S.; Smith, D.C. The effects of situational factors on in-store grocery shopping behavior: The role of store environment and time available for shopping. *J. Consum. Res.* **1989**, *15*, 422–433. [CrossRef]
42. Bialkova, S.; Grunert, K.G.; van Trijp, H. From desktop to supermarket shelf: Eye-tracking exploration on consumer attention and choice. *Food Qual. Prefer.* **2020**, *81*, 103839. [CrossRef]
43. Laguna, L.; Fiszman, S.; Puerta, P.; Chaya, C.; Tárrega, A. The impact of COVID-19 lockdown on food priorities Results from a preliminary study using social media and an online survey with Spanish consumers. *Food Qual. Prefer.* **2020**, *86*, 104028. [CrossRef] [PubMed]
44. Hobbs, J.E. Food supply chains during the COVID-19 pandemic. *Can. J. Agric. Econ.* **2020**. [CrossRef]
45. Yoon, J.; Narasimhan, R.; Kim, M.K. Retailer's sourcing strategy under consumer stockpiling in anticipation of supply disruptions. *Int. J. Prod. Res.* **2018**, *56*, 3615–3635. [CrossRef]
46. Frank, B.; Schvaneveldt, S.J. Understanding consumer reactions to product contamination risks after national disasters: The roles of knowledge, experience, and information sources. *J. Retail. Consum. Serv.* **2016**, *28*, 199–208. [CrossRef]
47. Asmundson, G.J.; Taylor, S. How health anxiety influences responses to viral outbreaks like COVID-19: What all decision-makers, health authorities, and health care professionals need to know. *J. Anxiety Disord.* **2020**, *71*, 102211. [CrossRef]
48. Sneath, J.Z.; Lacey, R.; Kennett-Hensel, P.A. Coping with a natural disaster: Losses, emotions, and impulsive and compulsive buying. *Mark. Lett.* **2009**, *20*, 45–60. [CrossRef]
49. Kennett-Hensel, P.A.; Sneath, J.Z.; Lacey, R. Liminality and consumption in the aftermath of a natural disaster. *J. Consum. Mark.* **2012**, *29*, 52–63. [CrossRef]
50. FAO. 2020. Available online: http://wwwfaoorg/2019-ncov/q-and-a/impact-on-food-and-agriculture/en/ (accessed on 10 July 2020).
51. Bernués, A.; Ripoll, G.; Panea, B. Consumer segmentation based on convenience orientation and attitudes towards quality attributes of lamb meat. *Food Qual. Prefer.* **2012**, *26*, 211–220. [CrossRef]
52. Realini, C.E.; Kallas, Z.; Pérez-Juan, M.; Gómez, I.; Olleta, J.L.; Beriain, M.J.; Albertí, P.; Sañudo, C. Relative importance of cues underlying Spanish consumers' beef choice and segmentation, and consumer liking of beef enriched with n-3 and CLA fatty acids. *Food Qual. Prefer.* **2014**, *33*, 74–85. [CrossRef]
53. Verain, M.C.; Sijtsema, S.J.; Antonides, G. Consumer segmentation based on food-category attribute importance: The relation with healthiness and sustainability perceptions. *Food Qual. Prefer.* **2016**, *48*, 99–106. [CrossRef]
54. Zeithaml, V.A. Consumer perceptions of price, quality and value: A means-end model and synthesis of evidence. *J. Mark.* **1988**, *52*, 2–22. [CrossRef]
55. Brunsø, K.; Fjord, T.A.; Grunert, K.G. *Consumers' Food Choice and Quality Perception*; The Aarhus School of Business Publ.: Aarhus, Denmark, 2002; pp. 1–60.

56. Lea, E.; Worsley, T. Australians' organic food beliefs, demographics and values. *Br. Food J.* **2005**, *107*, 855–869. [CrossRef]
57. Veale, R.; Quester, P. Consumer sensory evaluations of wine quality: The respective influence of price and country of origin. *J. Wine Econ.* **2008**, *3*, 10–29. [CrossRef]
58. Realini, C.E.; i Furnols, M.F.; Sañudo, C.; Montossi, F.; Oliver, M.A.; Guerrero, L. Spanish, French and British consumers' acceptability of Uruguayan beef, and consumers' beef choice associated with country of origin, finishing diet and meat price. *Meat Sci.* **2013**, *95*, 14–21. [CrossRef]
59. Utami, H.N.; Sadeli, A.H.; Perdana, T. Customer value creation of fresh tomatoes through branding and packaging as customer perceived quality. *J. ISSAAS* **2016**, *22*, 123–136.
60. Hu, L.; Baldin, A. The country of origin effect: A hedonic price analysis of the Chinese wine market. *Br. Food J.* **2018**, *120*, 1264–1279. [CrossRef]
61. Adasme-Berríos, C.; Sánchez, M.; Mora, M.; Díaz, J.; Schnettler, B.; Lobos, G. The gender role on moderator effect of food safety label between perceived quality and risk on fresh vegetables. *Rev. Fac. Cienc. Agrar. Univ. Cuyo* **2019**, *51*, 93–109.
62. Verbeke, W.; Ward, R.W. Consumer interest in information cues denoting quality, traceability and origin: An application of ordered probit models to beef labels. *Food Qual. Prefer.* **2006**, *17*, 453–467. [CrossRef]
63. Verbeke, W.; Roosen, J. Market differentiation potential of country-of-origin, quality and traceability labeling. *Estey Cent. J. Int. Law Trade Policy* **2009**, *10*, 20–35.
64. Bernabéu, R.; Díaz, M.; Olmeda, M. Origin vs organic in Manchego cheese: Which is more important? *Br. Food J.* **2010**, *112*, 887–901. [CrossRef]
65. Hersleth, M.; Næs, T.; Rødbotten, M.; Lind, V.; Monteleone, E. Lamb meat-Importance of origin and grazing system for Italian and Norwegian consumers. *Meat Sci.* **2012**, *90*, 899–907. [CrossRef]
66. Lim, K.H.; Hu, W.; Maynard, L.J.; Goddard, E. US consumers' preference and willingness to pay for country-of-origin-labeled beef steak and food safety enhancements. *Can. J. Agric. Econ.* **2013**, *61*, 93–118. [CrossRef]
67. Guzmán, A.C.; Schnettler, B.; Mora, M.; Aguilera, M. Perceived quality of and satisfaction from sweet cherries (*Prunus avium* L.) in China: Confirming relationships through structural equations. *J. Agric. Nat. Resour.* **2018**, *45*, 210–219. [CrossRef]
68. Salazar-Ordóñez, M.; Cordón-Pedregosa, R.; Rodríguez-Entrena, M.A. Consumer behavior approach to analyse handmade and locally made agrifood products in Western Honduras. *Econ. Agrar. Recur. Nat.* **2018**, *18*, 5–27. [CrossRef]
69. Glitsch, K. Consumer perceptions of fresh meat quality: Cross-national comparison. *Br. Food J.* **2000**, *102*, 177–194. [CrossRef]
70. Weatherell, C.; Tregear, A.; Allinson, J. In search of the concerned consumer: UK public perceptions of food, farming and buying local. *J. Rural Stud.* **2003**, *19*, 233–244. [CrossRef]
71. Bredahl, L. Cue utilisation and quality perception with regard to branded beef. *Food Qual. Prefer.* **2004**, *15*, 65–75. [CrossRef]
72. Verbeke, W.; Vackier, I. Profile and effects of consumer involvement in fresh meat. *Meat Sci.* **2004**, *67*, 159–168. [CrossRef] [PubMed]
73. Röhr, A.; Lüddecke, K.; Drusch, S.; Müller, M.J.; Alvensleben, R.V. Food quality and safety—Consumer perception and public health concern. *Food Control* **2005**, *16*, 649–655. [CrossRef]
74. Sepúlveda, W.; Maza, M.T.; Mantecón, A.R. Factors that affect and motivate the purchase of quality-labelled beef in Spain. *Meat Sci.* **2008**, *80*, 1282–1289. [CrossRef] [PubMed]
75. Martínez-Carrasco, L.; Brugarolas, M.; Martínez-Poveda, A.; Ruiz, J.J. Modelling Perceived Quality of Tomato by Structural Equation Analysis. *Br. Food J.* **2012**, *114*, 1414–1431. [CrossRef]
76. Papanagiotou, P.; Tzimitra-Kalogianni, I.; Melfou, K. Consumers' expected quality and intention to purchase high quality pork meat. *Meat Sci.* **2013**, *93*, 449–454. [CrossRef] [PubMed]
77. McDaniel, C.; Baker, R.C. Convenience food packaging and the perception of product quality. *J. Mark.* **1977**, *41*, 57. [CrossRef]
78. Peters-Texeira, A.; Badrie, N. Consumers' perception of food packaging in Trinidad, West Indies and its related impact on food choices. *Int. J. Consum. Stud.* **2005**, *29*, 508–514. [CrossRef]
79. Ampuero, O.; Vila, N. Consumer perceptions of product packaging. *J. Consum. Mark.* **2006**, *23*, 100–112. [CrossRef]

80. Hussain, S.; Ali, S.; Ibrahim, M.; Noreen, A.; Ahmad, S.F. Impact of product packaging on consumer perception and purchase intention. *J. Mark. Consum. Res.* **2011**, *10*, 1–10.
81. Magnier, L.; Schoormans, J.; Mugge, R. Judging a product by its cover: Packaging sustainability and perceptions of quality in food products. *Food Qual. Prefer.* **2016**, *53*, 132–142. [CrossRef]
82. Lo, S.C.; Tung, J.; Huang, K.P. Customer perception and preference on product packaging International. *J. Organ. Innov.* **2017**, *9*, 3–15. [CrossRef]
83. Wang, E.S.T. Different effects of utilitarian and hedonic benefits of retail food packaging on perceived product quality and purchase intention. *J. Food Prod. Mark.* **2017**, *23*, 239–250. [CrossRef]
84. Benachenhou, S.; Guerrich, B.; Moussaoui, Z. The effect of packaging elements on purchase intention: Case study of Algerian customers. *Manag. Sci. Lett.* **2018**, *8*, 217–224. [CrossRef]
85. Ahsan Ansari, M.U.; Siddiqui, D.A. Packaging features and consumer buying behavior towards packaged food items. *Glob. Sci. J.* **2019**, *7*, 1050–1073. [CrossRef]
86. Becker, T. Consumer perception of fresh meat quality: A framework for analysis. *Br. Food J.* **2000**, *102*, 158–176. [CrossRef]
87. Jover, A.J.V.; Montes, F.J.L.; Fuentes, M.D.M.F. Measuring perceptions of quality in food products: The case of red wine. *Food Qual. Prefer.* **2004**, *15*, 453–469. [CrossRef]
88. Grunert, K.G. Food quality and safety: Consumer perception and demand. *Eur. Rev. Agric. Econ.* **2005**, *32*, 369–391. [CrossRef]
89. Jaafar, S.N.; Lalp, P.E.; Naba, M.M. Consumers' perceptions, attitudes and purchase intention towards private label food products in Malaysia. *Asian J. Bus. Manag. Sci.* **2012**, *2*, 73–90.
90. Rubio, N.; Oubina, J.; Villasenor, N. Brand awareness–Brand quality inference and consumer's risk perception in store brands of food products. *Food Qual. Prefer.* **2014**, *32*, 289–298. [CrossRef]
91. Saleem, S.; Rahman, S.U.; Umar, R.M. Measuring customer based beverage brand equity: Investigating the relationship between perceived quality, brand awareness, brand image, and brand loyalty. *Int. J. Mark. Stud.* **2015**, *7*, 66. [CrossRef]
92. De Toni, D.; Eberle, L.; Larentis, F.; Milan, G.S. Antecedents of perceived value and repurchase intention of organic food. *J. Food Prod. Mark.* **2018**, *24*, 456–475. [CrossRef]
93. Susilowati, E.; Sari, A.N. The influence of brand awareness, brand association, and perceived quality toward consumers'purchase intention: A case of richeese factory, Jakarta. *Indep. J. Manag.* **2020**, *11*, 39–53. [CrossRef]
94. Larceneux, F.; Benoit-Moreau, F.; Renaudin, V. Why might organic labels fail to influence consumer choices? Marginal labelling and brand equity effects. *J. Consum. Policy* **2012**, *35*, 85–104. [CrossRef]
95. Lee, H.J.; Hwang, J. The driving role of consumers' perceived credence attributes in organic food purchase decisions: A comparison of two groups of consumers. *Food Qual. Prefer.* **2016**, *54*, 141–151. [CrossRef]
96. Konuk, F.A. The role of store image, perceived quality, trust and perceived value in predicting consumers' purchase intentions towards organic private label food. *J. Retail. Consum. Serv.* **2018**, *43*, 304–310. [CrossRef]
97. Curvelo, I.C.G.; de Morais Watanabe, E.A.; Alfinito, S. Purchase intention of organic food under the influence of attributes, consumer trust and perceived value. *Rev. Gestão* **2019**, *26*, 198–211. [CrossRef]
98. John, S.F.; Kumari, R.S.A. Study on Perceived knowledge towards Organic Products. *Int. J. Manag. IT Eng.* **2019**, *7*, 158–163.
99. Bonnet, C.; Simioni, M. Assessing consumer response to Protected Designation of Origin labelling: A mixed multinomial logit approach. *Eur. Rev. Agric. Econ.* **2001**, *28*, 433–449. [CrossRef]
100. Sepúlveda, W.S.; Maza, M.T.; Mantecón, Á.R. Factors associated with the purchase of designation of origin lamb meat. *Meat Sci.* **2010**, *85*, 167–173. [CrossRef]
101. i Furnols, M.F.; Realini, C.; Montossi, F.; Sañudo, C.; Campo, M.M.; Oliver, M.A.; Nute, G.R.; Guerrero, L. Consumer's purchasing intention for lamb meat affected by country of origin, feeding system and meat price: A conjoint study in Spain, France and United Kingdom. *Food Qual. Prefer.* **2011**, *22*, 443–451. [CrossRef]
102. Agnoli, L.; Capitello, R.; Begalli, D. Geographical brand and country-of-origin effects in the Chinese wine import market. *J. Brand Manag.* **2014**, *21*, 541–558. [CrossRef]
103. Dias, C.; Mendes, L. Protected designation of origin (PDO), protected geographical indication (PGI) and traditional speciality guaranteed (TSG): A bibliometric analysis. *Food Res. Int.* **2018**, *103*, 492–508. [CrossRef]
104. Lewis, K.E.; Grebitus, C.; Colson, G.; Hu, W. German and British consumer willingness to pay for beef labeled with food safety attributes. *J. Agric. Econ.* **2017**, *68*, 451–470. [CrossRef]

105. Lopes, C.; Leitão, J.; Rengifo-Gallego, J. Place branding: Revealing the neglected role of agro food products. *Int. Rev. Public Nonprofit Mark.* **2018**, *15*, 497–530. [CrossRef]
106. Rodríguez, S.E.; López, E.A.C. Los "mundos de producción" de las denominaciones de origen protegidas del vino en España: Disparidad de convenciones tecnológicas y comerciales. *Econ. Agrar. Recur. Nat.* **2017**, *17*, 101–125. [CrossRef]
107. Villanueva, J.L.J.; Lázaro, C.E.C.; Ávalos, V.C. Disponibilidad a pagar por atributos culturales en chocolates caseros artesanales de la región de la Chontalpa, Tabasco, México. *Econ. Agrar. Recur. Nat.* **2018**, *18*, 53–73. [CrossRef]
108. Wang, C.; Pan, R.; Wan, X.; Tan, Y.; Xu, L.; Ho, C.S.; Ho, R.C. Immediate psychological responses and associated factors during the initial stage of the 2019 coronavirus disease (COVID-19) epidemic among the general population in China. *Int. J. Environ. Res. Public Health* **2020**, *17*, 1729. [CrossRef] [PubMed]
109. Kee, H.T.; Wan, D. Intended Usage of Online Supermarkets: The Singapore Case, in ICEB 2004, 1308–1312. Available online: https://pdfssemanticscholarorg/6543/d5abae8db32b9d2975dc020e8bfbe4c1d9ce.pdf (accessed on 15 September 2020).
110. Forster, P.W.; Ya, T. The role of online shopping and fulfilment in the Hong Kong SARS crisis. In Proceedings of the 38th Annual Hawaii International Conference on System Sciences, Big Island, HI, USA, 6 January 2005. [CrossRef]
111. AECOC. Consumo y Compra Dentro y Fuera del Hogar Durante y Después del Covid-19. 2020. Available online: https://wwwaecoces/estudio/1er-informe-consumo-y-compra-dentro-y-fuera-del-hogar-durante-y-despues-del-covid-19/ (accessed on 15 July 2020).
112. Xie, X.; Huang, L.; Li, J.J.; Zhu, H. Generational Differences in Perceptions of Food Health/Risk and Attitudes toward Organic Food and Game Meat: The Case of the COVID-19 Crisis in China. *Int. J. Environ. Res. Public Health* **2020**, *17*, 3148. [CrossRef]

foods

MDPI

Article

Crocetin Isolated from the Natural Food Colorant Saffron Reduces Intracellular Fat in 3T3-L1 Adipocytes

Elena Jiménez-Ortega [1], Aitana Braza-Boïls [2], Miguel Burgos [3], Natalia Moratalla-López [4], Manuel Vicente [5], Gonzalo L. Alonso [4], Eduardo Nava [5] and Sílvia Llorens [5,*]

1 Department of Crystallography and Structural Biology, Institute of Physical-Chemistry Rocasolano, CSIC, 28006 Madrid, Spain; xelena@iqfr.csic.es

2 Unidad de Cardiopatías Familiares, Muerte Súbita y Mecanismos de Enfermedad (CaFaMuSMe), Instituto de Investigación Sanitaria La Fe, 46026 Valencia, Spain; aitana_braza@iislafe.es

3 Translational Research Unit, Albacete University Hospital, 02008 Albacete, Spain; miguel.burgos@uclm.es

4 Cátedra de Química Agrícola, ETSI Agrónomos y de Montes, Universidad de Castilla-La Mancha, Campus Universitario, 02071 Albacete, Spain; Natalia.Moratalla@uclm.es (N.M.-L.); Gonzalo.Alonso@uclm.es (G.L.A.)

5 Department of Medical Sciences, Faculty of Medicine of Albacete, Centro Regional de Investigaciones Biomédicas (CRIB), University of Castilla-La Mancha, 02008 Albacete, Spain; Manuel.Vicente@alu.uclm.es (M.V.); eduardo.nava@uclm.es (E.N.)

* Correspondence: silvia.llorens@uclm.es

Received: 7 October 2020; Accepted: 10 November 2020; Published: 12 November 2020

Abstract: Saffron, as a food colorant, has been displaced by low-cost synthetic dyes. These have unhealthy properties; thus, their replacement with natural food colorants is an emerging trend. Obesity is a worldwide health problem due to its associated comorbidities. Crocetin esters (crocins) are responsible for the red saffron color. Crocetin (CCT) exhibits healthful properties. We aimed to broaden the existing knowledge on the health properties of CCT isolated from saffron, to facilitate its consideration as a healthy natural food colorant in the future. We evaluated the ability of CCT (1 and 5 μM) to reduce lipid accumulation during the differentiation of 3T3-L1 preadipocytes. Intracellular fat was quantified by Oil Red O staining. CTT cytotoxicity was measured using the 3-(4,5-dimethylthiazol-2-yl)-2,5-diphenyltetrazolium bromide (MTT) assay. The number and size of lipid droplets were analyzed using WimLipid software. The expression of adipogenic genes (CCAAT/enhancer-binding protein (C/EBPβ, C/EBPδ, C/EBPα), and peroxisome proliferator-activated receptor γ (PPARγ)) was analyzed using quantitative real-time PCR (qRT-PCR). CCT 5 μM decreased intracellular fat by 22.6%, without affecting viability or lipid droplet generation, via a decrease in C/EBPα expression, implicated in lipid accumulation. Thus, CCT is a potential candidate to be included in dietary therapies aimed at reversing adipose tissue accumulation in obesity.

Keywords: saffron; crocetin; obesity

1. Introduction

Additive colorants are found in large quantities in food because consumers often associate them with the flavor, safety, and nutritional value of the foodstuff, thus making it more attractive [1]. The food industry uses synthetic food colorants to a large extent, mainly due to their low cost and high stability [2]; however, most of them contain azo functional groups and aromatic ring structures, which can be harmful to human health [3], with additional environmentally harmful effects during food processing [4]. Thus, an acceptable daily intake (ADI) of authorized food additives is continuously evaluated by regulatory bodies, such as the European Food Safety Authority (EFSA) in Europe and the

Food and Drug Administration (FDA) in the United States (US), to adequately protect consumers [5]. Even so, consumers remain cautious with regard to the safety of synthetic dyes, whereas they are also aware that many natural colorants provide health benefits; thus, the replacement of artificial food colorants with natural ones is a current market demand [2].

Naturally occurring color additives from vegetable and mineral sources were used to color foods, drugs, and cosmetics in ancient times. Paprika, turmeric, and saffron are some examples obtained from vegetables. From the second half of the 19th century onward, artificial colors rapidly replaced natural colorants. Artificial colorants are used in a wide variety of foods, mainly to make them more attractive to consumers; in fact, their intake by consumers has increased since 1950 [2]. Currently, consumers prefer natural colorants in foodstuff or for cookery since almost all of them are hypoallergenic and nontoxic, while displaying salutary properties in humans, which is a significant advantage over many artificial dyes. In addition, their production, use, and elimination is environmentally friendly and can contribute to sustainable development [6]. The current market trend involves the replacement of artificial dyes with natural ones [2].

The most important use of saffron (*Crocus sativus* L.) is in cookery; its dried stigmas constitute the saffron spice. This spice is greatly valued for its coloring, flavoring, and aromatizing properties in some traditional dishes, as well as in modern cuisine. In addition to its use as a spice, saffron has long been considered a medicinal plant for its therapeutic properties [7]. However, for years, the use of this spice as a food colorant, due to its high price, has been replaced by low-cost synthetic dyes (e.g., tartrazine (TTZ)) [8]. The consumption of TTZ can produce adverse metabolic effects [3,9–11]. Indeed, TTZ has been banned in some countries including Norway and Austria [12]. Turmeric (whose main coloring component is curcumin) has excellent heat stability, and it is often used as a replacement for TTZ; however, this pigment is unstable when exposed to light and it is susceptible to oxidation [2]. In contrast, saffron pigments are quite light- and heat-resistant [13].

Saffron contains several bioactive compounds, of which crocins (crocetin esters), a group of water-soluble carotenoids derived from crocetin (CCT, the aglycon of crocin), are responsible for the intense color that saffron provides to aqueous solutions [14]. Saffron displays numerous functional and bioactive properties. Therefore, research into the effects of saffron and its components is necessary to achieve a more widespread use of the spice. Today, along with the current trend of using natural colorants, there is a growing interest in therapeutic diets, which include culinary herbs or spices to support therapies for chronic diseases, including obesity [14]. Obesity is a global health problem that is acquiring an enormous epidemiological relevance due to its increasing prevalence rate [15]. The World Health Organization defines obesity as an abnormal or excessive accumulation of fat (adiposity) that can be harmful to health, and it is a risk factor for diabetes, cardiovascular disease, and cancer [16].

Recently, the most important therapeutic effects of saffron were attributed to CCT in its free-acid form [17]. Typical carotenoids contain 40 carbon atoms (C40); however, CCT is a C20 apocarotenoid ($C_{20}H_{24}O_4$; molecular weight 328.4 g/mol), and it is generated via the hydrolysis of crocin glycosides. Crocin is crocetin digentiobiase ester, whereas CCT is 8,8′-diapo-ψ,ψ′-carotenoic acid. CCT contains a carboxyl group at each end of the polyene chain; when ionized, it can function as an acid (anionic) dye for biological staining [18]. On the other hand, CCT has high antioxidant power and possesses a wide range of beneficial properties for humans including anti-inflammatory, antiatherosclerotic, antihypertensive, and anticancer activities [19–22].

Adipose tissue function is essential for health; it is pivotal in the synthesis and storage of triacylglycerol in lipid droplets (lipogenesis) and the release of fatty acids into systemic circulation during periods of scarcity. In addition, adipocytes are a source of numerous proteins and hormones with actions relevant in practically every aspect of human physiology, including cardiovascular physiology. A crucial process for the homeostatic maintenance of lipid metabolism is the generation of new adipocytes from preadipocytes, a process known as adipogenesis. The process of adipogenesis involves growth arrest, mitotic clonal expansion, early differentiation, and terminal differentiation [23]. In vitro, adipogenesis takes place in two sequential stages: (1) the early stage, dependent on the activation of early transcription factors: CCAAT/enhancer-binding protein (*C/EBP*)β and *C/EBP*δ, which in turn activate the transcription factors of the (2) late stage, dependent on the activation of late genes: *C/EBP*α and peroxisome proliferator-activated receptor γ (*PPAR*γ) [24–26]. In this way, preadipocytes differentiate into an adipocytic phenotype, causing morphological changes in the cell, including lipogenesis [27]. An exquisitely accurate adipogenesis process preserves lipid health [28], with increased lipid accumulation caused by an altered adipogenic process being a key factor in obesity. Thus, intervention in the regulation of adipogenesis, in terms of reducing fat mass, has been proposed as a possible therapy to prevent adipose tissue development and obesity [29]. In this sense, several studies have shown that CCT could play a preventive or even therapeutic role in some aspects related to the comorbidities that accompany obesity. Indeed, CCT has been shown to prevent visceral fat accumulation and insulin resistance induced by a hypercaloric diet in rats [30]. In addition, CCT regulates the expression of adiponectin in the adipose tissue of fructose-fed rats [31].

We observed in previous studies that different components of saffron, such as CCT and crocins, on some occasions, have opposite vasoactive properties [21]. Therefore, during this trend of a change toward healthier and more sustainable natural products, coinciding with the rapid advance of obesity in the world, we aimed to study CCT isolated from saffron (*C. sativus* L.) to broaden the existing knowledge on its beneficial properties and to promote its use as a healthy natural food colorant in the future. Specifically, our aim was to test the ability of CCT to reduce adipocytic lipid accumulation. We examined the ability of the CCT to induce differentiation in cultured murine 3T3-L1 preadipocytes by studying the amount of intracellular fat, the number and size of lipid droplets, and the viability and expression of the main early (*C/EBP*β and *C/EBP*δ) and late (*C/EBP*α and *PPAR*γ) genes involved in differentiation from preadipocytes to adipocytes. CCT decreased intracellular fat in mature adipocytes, showing potential antiadipogenic properties. Additionally, CCT did not affect lipid droplet generation or cellular viability. On the other hand, we report here that CCT diminished the messenger RNA (mRNA) levels of the transcription factor *C/EBP*α, which is implicated in lipid accumulation. Therefore, we propose that CCT reduces intracellular fat by decreasing *C/EBP*α mRNA levels.

2. Materials and Methods

2.1. Plant Material and Isolation of CCT

Saffron was obtained from the "Agrícola Técnica de Manipulación y Comercialización" company (Minaya, Albacete, Spain) during the 2014–2015 harvest. These dried stigmas belonged to the Protected Designation of Origin (PDO) "Azafrán de La Mancha", which complies with ISO 3632:2011 (Category I) and guarantees their origin and freedom from fraud. Saffron with a very low moisture level was stored in the dark at 4 °C until further use.

CCT was obtained via the hydrolysis of aqueous solutions of saffron acquired using a protected internal method of the "Verdú Cantó Saffron Spain" company (Novelda, Alicante, Spain) [22]. CCT purity was checked through the reverse-phase (RP)-HPLC–diode array detection (DAD) technique according to [32]. Twenty microliters of aqueous extracts of CCT were filtered through a syringe with a polytetrafluoroethylene (PTFE) filter, 0.45 µm pore size (Millipore, Bedford, MA, USA), and injected into an Agilent 1200 chromatograph (Palo Alto, CA, USA). Chromatographic determination was achieved using a Phenomenex Luna C18 column (150 × 4.6 mm, 5 µm) (Le Pecq CEDEX, France) equilibrated at 30 °C. Acetonitrile (ACN) and Milli Q water (mQW) were used as the mobile phase at a flow rate of 0.8 mL/min. HPLC-grade ACN was obtained from Panreac® (Barcelona, Spain) and ultrahigh-purity water mQW was produced using a Milli-Q system (Millipore, Danvers, MA, USA). The elution gradient was set up for the ACN solvent as follows: 20%, 0–5 min; 20–80%, 5–15 min; 80%, 15–18 min; and 20%, 18–30 min. The DAD detector (Hewlett Packard, Waldbronn, Germany) was set to 440 nm for *cis/trans*-CCT detection. The chromatographic purity of *cis/trans*-CCT according to HPLC–DAD at 440 nm was 99% (86% *trans*-CCT, retention time: 16.64 min; 13% *cis*-CCT, retention time:

17.86 min). CCT was stored at −20 °C until further use. Before using CCT to carry out the experiments, purity was determined again obtaining the same chromatographic purity before it was employed.

2.2. 3T3-L1 Cell Culture and Adipocyte Differentiation

The cell line of embryonic fibroblasts, 3T3-L1, was acquired in 2016 from the ATCC (American Type Culture Collection, Manassas, VA, USA). 3T3-L1 preadipocytes were cultured, maintained, and differentiated according to the supplier's instructions. In all experiments, cells were used within the sixth passage. Briefly, 3T3-L1 cells were expanded in a 75 cm^2 flask at 37 °C under a humidified 5% CO_2 atmosphere in preadipocyte expansion medium (EM; Dulbecco's modified Eagle's medium (DMEM, 90%) supplemented with L-glutamine (1%), penicillin/streptomycin (0.5%), and inactivated bovine calf serum (BCS, 10%). When the cells reached 70–80% confluence, they were seeded on six- or 96-well sterile plates and grown in EM for 48 h or until the culture reached 90% confluence. Then, differentiation of adipocytes was induced in the absence (as a control of differentiation) or presence of CCT (1 or 5 μM). Stimulating and inhibiting controls of differentiation were also established with rosiglitazone (10 μM, an agonist of PPARγ that activates adipogenesis [33,34]) or genistein (12.5 μM, an isoflavone that inhibits adipogenesis [35,36]), respectively. Genistein at this concentration inhibits lipid accumulation while preserving the viability of preadipocytes [36]. For this, post-confluent cells were treated for 48 h with differentiation medium (DM). The DM was prepared using the same components as the EM, instead replacing CBS with inactivated fetal bovine serum (FBS, 10%) and adding an adipogenic cocktail (AC), containing substances to induce differentiation (0.5 mM 3-isobutyl-1-methylxanthine (IBMX), 0.25 μM dexamethasone, and 1 μg/mL insulin (INS)). The DM was subsequently replaced with adipocyte maintenance medium (MM), which was composed of DMEM containing 1 μg/mL INS and 10% FBS. Cells were maintained in MM for 6 days, with the medium replenished every 2 days. At this point, the cells developed large lipid droplets and were considered mature adipocytes (Figure 1).

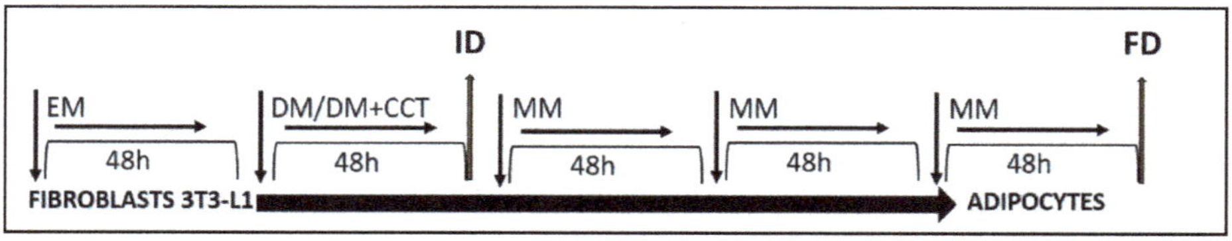

Figure 1. Schematic representation of the differentiation procedure to analyze the effects of crocetin (CCT). Fibroblasts achieved confluence after 48 h of treatment with expansion medium (EM). Then, the differentiation of confluent fibroblasts was induced with differentiation medium (DM; with or without CCT (1 or 5 μM)) for 48 h. Subsequently, preadipocytes were maintained with maintenance medium (MM) for 6 days. ID, initial differentiation; FD, final differentiation.

To check the effect of CCT on the early and late genes of differentiation, cells were collected at two time-points: 48 h after induction (initial differentiation, ID) and 6 days after differentiation (final differentiation, FD).

Two concentrations of CCT (1 and 5 μM) were tested in this work. This choice was based on our experience in previous studies [22] and on the work of Chryssanthi et al. [37]. These were prepared from a stock CCT solution, dissolved in sterile dimethyl sulfoxide (DMSO) and added to DM (the final concentration of DMSO in the culture medium was 0.001% (*v/v*)). Rosiglitazone and genistein stock solutions were also prepared in DMSO and added to the DM to reach a working concentration (the final volume of DMSO in the well was 0.001% (*v/v*)). The differentiation of adipocytes was also carried out in the presence of sterile DMSO (0.001%) as the solvent control. DMSO is usually well tolerated with no observable toxic effects on cells at a 0.1% final concentration. This compound is widely used as a solvent for various pharmacological agents at concentrations of 0.05–1.5% [38].

2.3. Quantification of the Intracellular Fat by Oil Red O Staining

Cells differentiated in 96-well sterile plates (cellular density, 4×10^3 cell/well) were stained with Oil Red O (OR) at the FD time-point, according to the method developed by Kraus and colleagues [39] with slight modification. Oil Red O is a dye that strongly stains lipids, specifically triacylglycerol, often used for the quantitative analysis of adipocyte differentiation [39].

The OR stock solution was prepared the day before as follows: 0.2 g OR was dissolved in 100 mL of isopropanol for 24 h at room temperature under agitation. The OR working solution was prepared by mixing six parts of OR stock solution and four parts of double-distilled water (ddH_2O). The solution was filtered through a two-layer Whatman paper to remove any precipitate.

To stain cells with the OR working solution, they were first washed three times with phosphate-buffered saline (PBS) and fixed in 4% formaldehyde for 1 h at room temperature while avoiding any shaking of the plate. Formaldehyde was removed, and the cells were washed once with cold PBS and air-dried for 10 min. The freshly prepared OR working solution was added to the plates to cover the cell surface. After 10 min, the solution was aspirated, and the cells were washed three times with cold PBS and air-dried for 15 min. OR was eluted with 100% isopropanol for 10 min, and absorbance was measured using a spectrophotometer (ASYS UVM 340, Cambridge, United Kingdom, Microplate Readers) at 450 nm. Data obtained from at least 10 replicates of each condition from three independent experiments were used for analysis. The amount of color produced is directly proportional to the amount of intracellular fat.

Reagents, unless specified otherwise, were acquired from Sigma-Aldrich.

2.4. Determination of the Number and Size of Lipid Droplets

The number and size distribution of lipid droplets were evaluated by Wimasis (Edificio Centauro, 14014 Córdoba, Spain) using a WimLipid image analysis software. For this, photomicrographs (20X) of the wells at the FD time-point were taken using a phase-contrast microscope (Olympus 1X51). Parameters such as the circularity, convexity, and elongation were included in the analysis to discriminate drops. The following criteria were used: area $\geq$ 10 pixels (Px); circularity > elongation; convexity > 0.95. Drops that failed to meet these criteria were removed.

2.5. Quantification of Cellular Viability

2.5.1. 3-(4,5-Dimethylthiazol-2-yl)-2,5-diphenyltetrazolium bromide (MTT) Assay

To perform the 3-(4,5-dimethylthiazol-2-yl)-2,5-diphenyltetrazolium bromide (MTT) assay, cells were grown on 96-well sterile plates at a cellular density of 4×10^3 cell/well. The MTT viability assay was carried out as previously described [40] with slight modification. The MTT assay measures mitochondrial activity in metabolizing cells and, therefore, can be used as an approximate measurement of cell viability. The assay relies on the reduction of MTT, a yellow water-soluble tetrazolium dye, primarily by mitochondrial dehydrogenases, to purple-colored formazan crystals. An MTT stock solution in PBS was freshly prepared and assessed in all experimental conditions. At the FD time-point, cells were washed with red phenol-free DMEM without FBS. Then, 100 μL of MTT solution (red phenol-free DMEM with MTT 0.5 μg/μL) was added to each well, mixed gently, and incubated for 45 min at 37 °C. Media were immediately aspirated and discarded; then, in order to solubilize formazan crystals, 100 μL of DMSO was added to each well before gently stirring for 3–5 min. Absorbance was determined spectrophotometrically at 570 nm using a reference wavelength of 630 nm (ASYS UVM 340, Cambridge, United Kingdom, Microplate Readers). The color intensity is directly proportional to the number of viable cells. Data obtained from at least 10 replicates in each experimental condition from three independent experiments were used for analysis. Absorbance was measured for wells containing the control differentiated cells (DM-differentiated cells) and the DM + CCT (1 or 5 μM)-differentiated cells. Furthermore, differentiated cells in the presence of solvent (DM + DMSO (0.001%)) and an activator (DM + rosiglitazone 10 μM) or inhibitor (DM + genistein 12.5 μM) of adipogenesis were also measured.

2.5.2. Trypan Blue Assay

The MTT assay detects viable cells but does not take into consideration cell loss caused by cell death; thus, the percentage of viable cells was additionally determined using a dye exclusion test with trypan blue (TB) dye, which is based on the principle that living cells possess intact cell membranes that exclude TB, whereas dead cells do not. The TB test was performed as previously described [41]. Briefly, an aliquot of cell suspension was centrifuged (950 rpm during 5 min, Eppendorf Centrifuge 5804, Hamburg, Germany), and the pellet was resuspended in PBS or serum-free complete medium. Then, 10 μL of cell suspension corresponding to each experimental condition was mixed with 10 μL of TB (0.4%). After incubating the mixture for 3 min at room temperature, the cells were examined using an automated cell counter (TC10TM, Hercules, CA, USA, BioRad).

2.6. Expression of Main Genes Related to Early and Late Differentiation

To analyze the expression of early and late genes, cells were collected at the ID and FD time-points, respectively. Total RNA was extracted from cells differentiated in sterile six-well plates (at a cellular density of 8 × 10^4 cell/well) using the extraction kit PureLink™ RNA Mini Kit (Waltham, MA USA, ThermoFisher Scientific), according to the manufacturer's instructions. The extracted RNA was verified and quantified spectrophotometrically using NanoDrop (Thermo Scientific). Complementary DNA (cDNA) was synthesized from 1 μg of RNA using the RevertAid H Minus First-Strand cDNA synthesis kit (Fisher Scientific), according to the manufacturer's protocol. Gene expression was assessed using quantitative real-time PCR (qRT-PCR) in a LightCycler 480 II thermocycler with Fast Sybr Green Master Mix (Waltham, MA USA, Applied Biosystem). β-Actin was used as an endogenous control. The primer sequences used for amplification are presented in Table A1 (Appendix A).

The reaction mixtures were incubated for an initial denaturation at 95 °C for 10 min, followed by 45 PCR cycles (95 °C for 15 s, 60 °C for 1 min, 95 °C for 15 s, and 60 °C for 1 min). The ΔΔCT method was used to measure relative quantification, and the levels of transcripts were normalized to that of β-actin. The levels of each mRNA were calculated as relative expression to the basal condition (designated as 1). 3T3-L1 cells treated for 48 h with the AC, without CCT, were considered representative of the basal condition. Three independent experiments were performed, each in duplicate.

Calibrated ΔCt values from undifferentiated and control differentiated cells were used to evaluate the expression of genes. Fold changes in gene expression were calculated using the $2^{-\Delta\Delta Ct}$ method [28]. Expression of the *aP2-1* gene was used as a specific adipocyte marker. *aP2* is widely used as a marker of differentiated adipocytes [42].

2.7. Data Analysis

All data were presented as the mean ± SD. One-way ANOVA and a post hoc Bonferroni's multiple-comparison test, using GraphPad Prism version 5.0 software, were used to identify differences between groups. The results were considered to be significant at a p-value < 0.05.

3. Results

3.1. Crocetin Reduced Intracellular Fat in 3T3-L1 Adipocytes

Oil Red O (OR) staining at the FD time-point allowed visualizing the effect of CCT on the storage of intracellular lipid in differentiated 3T3-L1 adipocytes. OR staining allows estimating the amount of intracellular fat. The maximum intracellular fat detected in the control differentiated cells (CCT 0 μM) at the FD time-point was set to 100%, and the relative intracellular fat levels at the two CCT concentrations at the same time-point are depicted in Figure 2. As shown in this figure, CCT-differentiated cells at 5 μM showed a significant decrease in the content of intracellular fat (77.4 ± 11.2%, $p < 0.01$) compared to control differentiated cells. No significant effect was observed on the content of intracellular fat in cells differentiated with 1 μM CCT (97.4 ± 5.4%) or with DMSO (101 ± 10%), with respect to control differentiated cells. As expected, rosiglitazone significantly enhanced intracellular fat (219 ± 42%) as compared to control and CCT-differentiated cells ($p < 0.001$), whereas genistein significantly reduced intracellular fat (78 ± 7%) as compared to the control ($p < 0.05$). The effect of 5 μM CCT did not present a significant difference compared to genistein.

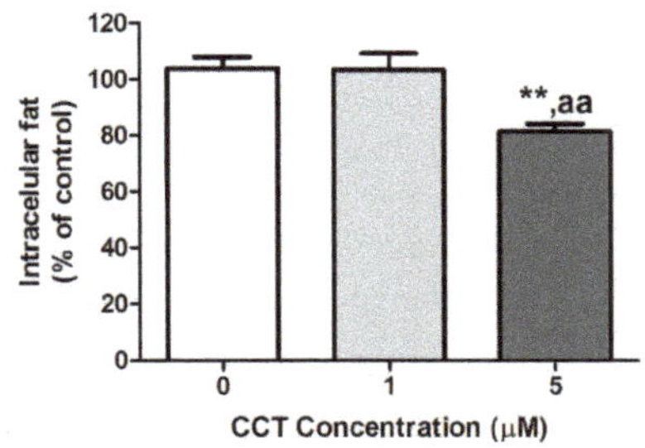

Figure 2. Effect of CCT on lipid accumulation. The maximum intracellular fat detected in control differentiated cells (CCT 0 μM) was set to 100%; the relative intracellular fat levels of CCT-differentiated cells are shown. Values were obtained at the FD time-point. Data from at least 10 replicates in each condition from three independent experiments are expressed as the mean ± SD. ** indicates a significant difference compared with the control differentiated cells ($p < 0.01$); ^{aa} indicates a significant difference compared with 1 μM CCT ($p < 0.01$).

3.2. *Crocetine Did Not Affect the Total Number of Lipid Drops or Their Size*

Additionally, we evaluated whether the presence of CCT during the induction of differentiation resulted in a reduction in the number or size of lipid droplets, which could explain the lower lipid load seen in Figure 2. This was analyzed at the FD time-point using the aforementioned WinLipid software. It was observed that the induction of differentiation in the presence of the two CCT concentrations did not affect the total number of lipid droplets (Figure 3A). Regarding the size of the lipid droplets, there was no statistical difference after the induction of differentiation in the presence of CCT; however, a slight shift to the left was observed in the size distribution frequency curve. Thus, after the induction of differentiation in the presence of CCT, the highest droplet percentage was found within a smaller interval size (60–109 Px) than that of the control differentiated cells (110–159 Px) (Figure 3B).

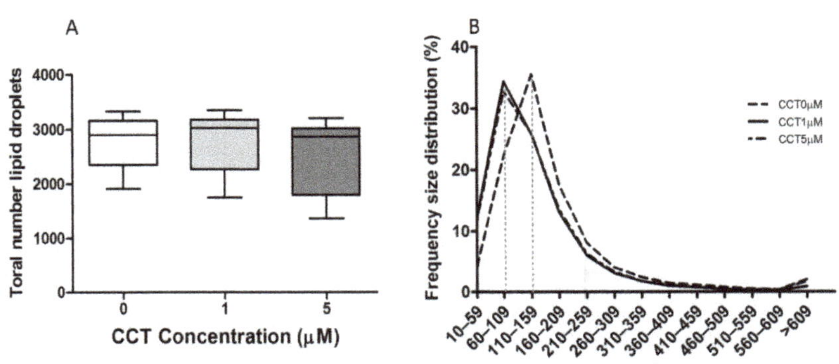

Figure 3. Effect of CCT on the number and size of lipid droplets. (**A**) Final lipid droplet number. Box-and-whisker plot depicting the minimum to maximum values, with the middle line representing the mean ± SD. (**B**) Intervals of frequency–area distribution. The vertical dotted line indicates the size

range in which most lipid droplets were found. Droplets were analyzed at the FD time-point using WinLipid software with 1–2 photomicrographs per well from three independent experiments; 0 μM CCT (control; n = 15), 1 μM CCT (n = 11), 5 μM CCT (n = 10), where n represents the number of analyzed photomicrographs. Px, pixels.

Thus, this result indicates that the reduction in intracellular fat produced by CCT was neither due to a lower generation of lipid droplets nor a reduction in their size.

3.3. Crocetin Did Not Affect 3T3-L1 Cell Viability

In order to assess the safety of CCT, two cell viability assays were conducted at the FD time-point. Figure 4A displays the percentage cellular metabolic activity of the 3T3-L1 adipocytes obtained using the MTT assay. This assay uses metabolic activity as an indicator of cell viability by evaluating the efficiency of mitochondrial enzymes. Mitochondrial activity generates a change in color measured by spectrophotometry. The amount of color produced is proportional to the metabolic activity, thus estimating cell viability. Our results indicate no statistically significant difference in cellular viability between the control differentiated cells (CCT 0 μM) and CCT-differentiated adipocytes. Therefore, the viability of 3T3-L1 adipocytes was not affected by the concentration of CCT (1 or 5 μM: 107 ± 8% or 97 ± 8%, respectively) or DMSO (106 ± 17%). Additionally, the percentage of viable cells was determined using TB dye. Figure 4B displays the percentage of live cells with respect to total cells. Similarly, to the results obtained with MTT, no statistically significant difference in the percentage of living cells was found between the control differentiated cells (CCT 0 μM) and CCT-differentiated adipocytes. Thus, the viability of 3T3-L1 adipocytes was not affected by the concentration of CCT (1 or 5 μM: 90 ± 13% or 103 ± 13%, respectively) or DMSO (95 ± 12%); furthermore, their viability was neither affected by rosiglitazone (111 ± 8%) nor genistein (102 ± 8%).

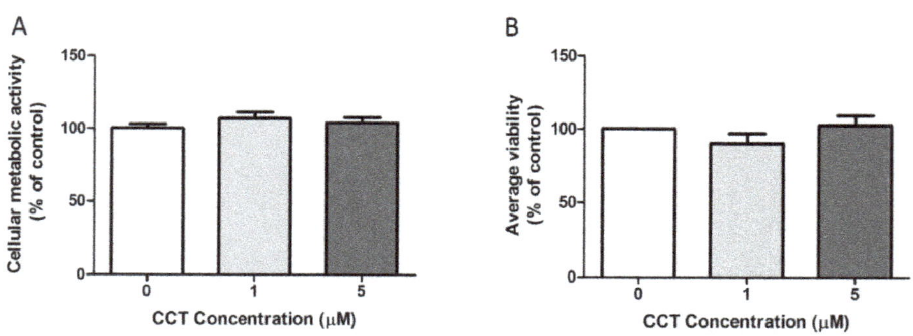

Figure 4. Effect of CCT on cell viability. The maximum viability detected for control differentiated cells (CCT 0 μM) was set to 100%; the relative viability of CCT-differentiated cells is shown. Values were obtained at the FD time-point. Data are expressed as the mean ± SD. (**A**) 3-(4,5-Dimethylthiazol-2-yl)-2,5-diphenyltetrazolium bromide (MTT) assay: data were obtained from at least 10 replicates in each condition from three independent experiments. (**B**) Trypan blue (TB) assay: two readings were performed per well using an automated cell counter. The percentage of viable cells (alive cells with respect to total cells) was averaged. Data were obtained from at least two replicates in each condition from three independent experiments.

In summary, our results indicate that neither CCT nor DMSO decreased the adipocyte mass due to cell death; thus, the decrease in intracellular fat observed in CCT-differentiated adipocytes was independent of nonspecific cell toxicity.

3.4. Crocetin Altered the Expression of Early and Late Genes during the Adipogenic Process

To determine the effect of CCT on the induction of the adipogenic process at a molecular level, the expression of early (*C/EBPβ* and *C/EBPδ*) and late (*PPARγ* and *C/EBPα*) genes of adipogenesis was quantified using qRT-PCR at the ID and FD time-points, respectively.

As shown in Figure 5 (panel A), 5 μM CCT significantly decreased *C/EBPβ* mRNA levels by 48.2% compared to control differentiated cells; however, this concentration of CCT did not alter *C/EBPδ* mRNA levels. On the other hand, 1 μM CCT modified the mRNA levels of both genes. As for the late genes, both doses of CCT (1 and 5 μM) significantly decreased *C/EBPα* mRNA levels (by 36.4% and 35.7%, respectively), without affecting the levels of *PPARγ* (Figure 5, panel B). The expression of *aP2-1* as a specific adipocyte marker was also evaluated at the FD time-point, showing no statistical difference in mRNA level with respect to control differentiated cells (Figure 5, panel B).

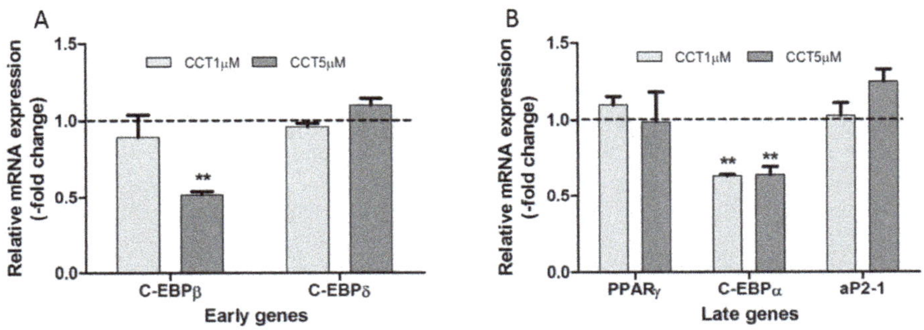

Figure 5. Effect of CCT on messenger RNA (mRNA) expression levels of genes regulating the adipogenic process. The relative quantitative real-time PCR (qRT-PCR) values were corrected to actin expression levels and normalized to control differentiation (0 μM CCT). Data were obtained from three independent experiments and are expressed as the mean ± SD. The maximum mRNA expression level for 0 μM CCT was set to 1 (dashed line); the relative mRNA expression levels obtained with the two CCT concentrations (1 and 5 μM) at the same time-point are depicted. (**A**) Cells were collected at the ID-time-point for the expression of early genes, CCAAT/enhancer-binding protein (C/EBP)β and *C/EBPδ*, evaluated using qRT-PCR with specific primer pairs. (**B**) Cells were collected at the FD time-point for the expression of late genes, peroxisome proliferator-activated receptor γ (*PPARγ*) and *C/EBPα*, evaluated using qRT-PCR with specific primer pairs; the expression of *aP2-1* (as a specific adipocyte marker) was also evaluated. ** indicates a significant difference compared with control differentiated cells ($p < 0.01$).

4. Discussion

In the present study, we showed that CCT could dose-dependently exert an antiadipogenic effect by decreasing lipid accumulation, a hallmark of antiobesity action. Our results showed that 5 μM CCT, but not 1 μM CCT, added at the early stage of cell differentiation reduced lipid accumulation. CCT exhibited the same efficacy as genistein in reducing intracellular fat. However, this result did not allow discriminating whether this decrease was the result of a lower generation of lipid droplets and, hence, a lower lipid load or if it was a cytotoxic effect of CCT on adipocytes via cell death, leading to a reduction in adipocyte mass and intracellular fat. To rule this out, lipid droplet metrics and cell viability were evaluated, with no differences observed in either analysis in the presence or absence of CCT.

Thus, we also investigated the effects of CCT on the expression of essential adipogenesis-related transcription factors involved in coordinating the adipogenic process. We analyzed the effect of CCT on the expression of early (*C/EBPβ* and *C/EBPδ*) and late (*C/EBPα* and *PPARγ*) genes which play key roles in adipogenesis. In this context, regulating the expression of these factors can modulate the differentiation capacity of adipocytes [43].

Typically, the mRNA levels of *C/EBPβ* and *C/EBPδ* increase during early differentiation, subsequently declining after the removal of adipogenic cocktail (AC), at which point the levels of *C/EBPα* and *PPARγ* increase [44,45]. It is known that *C/EBPβ* and *C/EBPδ* promote adipogenesis, at least in part, by inducing *C/EBPα* and *PPARγ*.

Interestingly, our results, at the ID time-point, indicate that adipocyte differentiation in the presence of the highest concentration of CCT led to a decrease in *C/EBPβ* mRNA levels, while those of C/EBPδ were not affected. However, at the FD time-point of differentiation, both tested concentrations of CCT diminished *C/EBPα* mRNA levels, while those of *PPARγ* remained unchanged.

The antiadipogenic effect of 5 μM is in line with reports showing that a knockdown of *C/EBPβ* inhibited adipocyte differentiation in 3T3-L1 preadipocytes [45]; however, in our work, CCT did not affect *C/EBPδ* expression. It is possible that an unaltered *C/EBPδ* expression can compensate for the decrease in C/EBPβ at the early stages of differentiation [46], thus playing a role in the induction of *PPARγ* expression. Most carotenoids inhibit the adipogenic process via the repression of *PPARγ* [47,48]; however, CCT did not alter *PPARγ* mRNA levels at the FD time-point. In this sense, as mentioned in Section 1, CCT is unlike β-carotene and other carotenoids, potentially exerting its effects via another mechanism [49]. On this note, it is known that β-carotene added during the differentiation of NIH 3T3-L1 preadipocytes reduced the triacylglycerol content, as well as the number and size of lipid droplets, compared with control differentiated cells by diminishing *PPARγ* expression [50]. On the other hand, Gul T et al. [51] reported that crocins at a concentration of 30 μM (treatment for 48 h) were able to diminish intracellular fat in 3T3-L1 cells, albeit producing a decrease in viability. PPARγ is not only crucial for adipogenesis but is also required for the maintenance of differentiated adipocytes [52]. It is known that knockdown of *PPARγ* compromises adipose tissue function, accompanied by insulin resistance, inflammation, angiogenesis, and fibrosis [53]. Thus, it appears that, by not altering *PPARγ* mRNA levels, CCT guarantees the culmination of the preadipocyte differentiation.

The decrease in *C/EBPβ* mRNA level seems to be CCT-dose-dependent, in contrast to that of *C/EBPα*. This suggests that CCT is an efficient carotenoid in that it selectively downregulates *C/EBPα* mRNA levels with no change in *PPARγ* expression. It has been reported that mice adipocytes in which the *C/EBPα* gene was disrupted showed defects in lipid accumulation [54]. Furthermore, various bioactive compounds reduce lipid accumulation in adipocytes by downregulating the expression of *C/EBPα* [55]. In line with these observations, the inhibition of *C/EBPα* reported herein could be involved in the decrease in intracellular fat caused by CCT during the induction of differentiation.

Taking our findings into consideration, further research should be conducted in this regard to determine the exact mechanism of CCT in combination with other standard therapeutic approaches applied to the obese population. This could contribute to the development of new strategies to improve the treatment of obesity.

Through our study, we have broadened the knowledge related to the health properties of CCT isolated from saffron as an antiadipogenic compound. Our research provides evidence that CCT efficiently reduces lipid accumulation in adipocytes, presumably via downregulation of *C/EBPα* expression. We can conclude that CCT is more potent than genistein in reducing lipid accumulation, with a similar effect produced at a lower concentration. It has been reported that genistein has a therapeutic effect on obesity [56], whereas several studies showed that its administration may be effective for adipocyte differentiation [57]. Thus, CCT's greater efficacy than genistein suggests that this natural compound is a potential candidate to be included in dietary therapies aimed at reverting adipose tissue accumulation in obesity.

Author Contributions: Conceptualization, S.L.; methodology, N.M.-L., A.B.-B., S.L., M.V. and E.J.-O.; validation, E.J.-O., A.B.-B. and S.L.; formal analysis, S.L.; investigation, S.L., E.J.-O., A.B.-B., E.N. and M.B.; resources, S.L., E.N., G.L.A., N.M.-L. and A.B.-B.; data curation, S.L. and E.J.-O.; writing—original draft preparation, S.L.; writing—review and editing, E.N., A.B.-B. and S.L.; visualization, S.L.; supervision, S.L.; project administration, S.L. All authors read and agreed to the published version of the manuscript.

Funding: This research received no external funding.

Acknowledgments: We acknowledge the administrative and technical support of Maria Carmen Cifuentes.

Conflicts of Interest: The authors declare no conflict of interest.

Appendix A

Table A1. Mouse primer sequences (5′-3′) used for amplification.

	Forward	Reverse
β-Actin	AGGGAAATCGTGCGTGACAT	GGAAAAGAGCCTCAGGGCAT
aP2	TCACCTGGAAGACAGCTCCT	AATCCCCATTTACGCTGATG
C/EBPβ	TTATAAACCTCCCGCTCGGC	CTCAGCTTGTCCACCGTCTT
C/EBPδ	AGAACCCGCGGCCTTCTAC	GTCGTACATGGCAGGAGTCG
C/EBPα	CCCTTGCTTTTTGCACCTCC	TGCCCCCATTCTCCATGAAC
PPARγ	CCAGAGTCTGCTGATCTGCG-	GCCACCTCTTTGCTCTGCTC

References

1. Clydesdale, F.M. Color as a factor in food choice. *Crit. Rev. Food Sci. Nutr.* **1993**, *33*, 83–101. [CrossRef] [PubMed]
2. Wrolstad, R.E.; Culver, C.A. Alternatives to those artificial FD&C food colorants. *Annu. Rev. Food Sci. Technol.* **2012**, *3*, 59–77. [CrossRef] [PubMed]
3. Amin, K.A.; Abdel Hameid, H.; Abd Elsttar, A.H. Effect of food azo dyes tartrazine and carmoisine on biochemical parameters related to renal, hepatic function and oxidative stress biomarkers in young male rats. *Food Chem. Toxicol.* **2010**, *48*, 2994–2999. [CrossRef] [PubMed]
4. Moutaouakkil, A.; Zeroual, Y.; Zohra Dzayri, F.; Talbi, M.; Lee, K.; Blaghen, M. Purification and partial characterization of azoreductase from Enterobacter agglomerans. *Arch. Biochem. Biophys.* **2003**, *413*, 139–146. [CrossRef]
5. EFSA Panel on Food Additives and Nutrient Sources Added to Food. Scientific Opinion on the re-evaluation Tartrazine (E 102). *EFSA J.* **2009**, *7*, 1331–1383. [CrossRef]
6. Nowik, W. DYES|Liquid Chromatography. In *Encyclopedia of Separation Science*; Wilson, I.D., Ed.; Academic Press: Oxford, UK, 2000; pp. 2602–2618. [CrossRef]
7. Ulbricht, C.; Conquer, J.; Costa, D.; Hollands, W.; Iannuzzi, C.; Isaac, R.; Jordan, J.K.; Ledesma, N.; Ostroff, C.; Serrano, J.M.; et al. An evidence-based systematic review of saffron (*Crocus sativus*) by the Natural Standard Research Collaboration. *J. Diet. Suppl.* **2011**, *8*, 58–114. [CrossRef]
8. Mehedi, N.; Ainad-Tabet, S.; Mokrane, N.; Addou, S.; Zaoui, C.; Kheroua, O.; Saidi, D. Reproductive Toxicology of Tartrazine (FD and C Yellow No. 5) in Swiss Albino Mice. *Am. J. Pharmacol. Toxicol.* **2009**, *4*, 130–135. [CrossRef]
9. Stevens, L.J.; Kuczek, T.; Burgess, J.R.; Stochelski, M.A.; Arnold, L.E.; Galland, L. Mechanisms of behavioral, atopic, and other reactions to artificial food colors in children. *Nutr. Rev.* **2013**, *71*, 268–281. [CrossRef]
10. Corkey, B.E.; Shirihai, O. Metabolic master regulators: Sharing information among multiple systems. *Trends Endrocrinol. Metab.* **2012**, *23*, 594–601. [CrossRef]
11. Axon, A.; May, F.E.B.; Gaughan, L.E.; Williams, F.M.; Blain, P.G.; Wright, M.C. Tartrazine and sunset yellow are xenoestrogens in a new screening assay to identify modulators of human oestrogen receptor transcriptional activity. *Toxicology* **2012**, *298*, 40–51. [CrossRef]
12. Guyot, S.; Serrand, S.; Le Quéré, J.M.; Sanoner, P.; Renard, C.M.G.C. Enzymatic synthesis and physicochemical characterisation of phloridzin oxidation products (POP), a new water-soluble yellow dye deriving from apple. *Innov. Food Sci. Emerg. Technol.* **2007**, *8*, 443–450. [CrossRef]
13. Mortensen, A. Carotenoids and other pigments as natural colorant. *Pure Appl. Chem.* **2006**, *78*, 1477–1491. [CrossRef]

14. Bagur, M.J.; Alonso Salinas, G.L.; Jiménez-Monreal, A.M.; Chaouqi, S.; Llorens, S.; Martínez-Tomé, M.; Alonso, G.L. Saffron: An Old Medicinal Plant and a Potential Novel Functional Food. *Molecules* **2017**, *23*, 30. [CrossRef] [PubMed]
15. Zimmet, P.; Alberti, K.G.; Shaw, J. Global and societal implications of the diabetes epidemic. *Nature* **2001**, *414*, 782–787. [CrossRef]
16. Sethi, J.K.; Vidal-Puig, A.J. Thematic review series: Adipocyte biology. Adipose tissue function and plasticity orchestrate nutritional adaptation. *J. Lipid Res.* **2007**, *48*, 1253–1262. [CrossRef] [PubMed]
17. Hashemi, M.; Hosseinzadeh, H. A comprehensive review on biological activities and toxicology of crocetin. *Food Chem. Toxicol.* **2019**, *130*, 44–60. [CrossRef] [PubMed]
18. Lillie, R.D. Harold Joel Conn 1886-1975. *Stain Technol.* **1977**, *52*, 3–4. [CrossRef] [PubMed]
19. Giaccio, M. Crocetin from saffron: An active component of an ancient spice. *Crit. Rev. Food Sci. Nutr.* **2004**, *44*, 155–172. [CrossRef]
20. Festuccia, C.; Mancini, A.; Gravina, G.L.; Scarsella, L.; Llorens, S.; Alonso, G.L.; Tatone, C.; Di Cesare, E.; Jannini, E.A.; Lenzi, A.; et al. Antitumor Effects of Saffron-Derived Carotenoids in Prostate Cancer Cell Models. *BioMed Res. Int.* **2014**, *2014*, 135048. [CrossRef]
21. Llorens, S.; Mancini, A.; Serrano-Díaz, J.; D'Alessandro, A.M.; Nava, E.; Alonso, G.L.; Carmona, M. Effects of Crocetin Esters and Crocetin from Crocus sativus L. on Aortic Contractility in Rat Genetic Hypertension. *Molecules* **2015**, *20*, 17570–17584. [CrossRef]
22. Mancini, A.; Serrano-Díaz, J.; Nava, E.; D'Alessandro, A.M.; Alonso, G.L.; Carmona, M.; Llorens, S. Crocetin, a Carotenoid Derived from Saffron (*Crocus sativus* L.), Improves Acetylcholine-Induced Vascular Relaxation in Hypertension. *J. Vasc. Res.* **2014**, *51*, 393–404. [CrossRef] [PubMed]
23. Gregoire, F.M.; Smas, C.M.; Sul, H.S. Understanding Adipocyte Differentiation. *Physiol. Rev.* **1998**, *78*, 783–809. [CrossRef] [PubMed]
24. Payne, V.A.; Au, W.S.; Lowe, C.E.; Rahman, S.M.; Friedman, J.E.; O'Rahilly, S.; Rochford, J.J. C/EBP transcription factors regulate SREBP1c gene expression during adipogenesis. *Biochem. J.* **2009**, *425*, 215–223. [CrossRef] [PubMed]
25. Rosen, E.D.; MacDougald, O.A. Adipocyte differentiation from the inside out. *Nat. Rev. Mol. Cell Biol.* **2006**, *7*, 885–896. [CrossRef]
26. White, U.A.; Stephens, J.M. Transcriptional factors that promote formation of white adipose tissue. *Mol. Cell Endocrinol.* **2010**, *318*, 10–14. [CrossRef]
27. Palomer, X.; Pérez, A.; Blanco-Vaca, F. Adiponectina: Un nuevo nexo entre obesidad, resistencia a la insulina y enfermedad cardiovascular. *Med. Clin.* **2005**, *124*, 388–395. [CrossRef]
28. Livak, K.J.; Schmittgen, T.D. Analysis of relative gene expression data using real-time quantitative PCR and the 2(-Delta Delta C(T)) Method. *Methods* **2001**, *25*, 402–408. [CrossRef]
29. McGregor, R.A.; Choi, M.S. microRNAs in the regulation of adipogenesis and obesity. *Curr. Mol. Med.* **2011**, *11*, 304–316. [CrossRef]
30. Sheng, L.; Qian, Z.; Shi, Y.; Yang, L.; Xi, L.; Zhao, B.; Xu, X.; Ji, H. Crocetin improves the insulin resistance induced by high-fat diet in rats. *Br. J. Pharmacol.* **2008**, *154*, 1016–1024. [CrossRef]
31. Xi, L.; Qian, Z.; Xu, G.; Zhou, C.; Sun, S. Crocetin attenuates palmitate-induced insulin insensitivity and disordered tumor necrosis factor-alpha and adiponectin expression in rat adipocytes. *Br. J. Pharmacol.* **2007**, *151*, 610–617. [CrossRef]
32. Valle García-Rodríguez, M.; Serrano-Díaz, J.; Tarantilis, P.A.; López-Córcoles, H.; Carmona, M.; Alonso, G.L. Determination of saffron quality by high-performance liquid chromatography. *J. Agric. Food Chem.* **2014**, *62*, 8068–8074. [CrossRef] [PubMed]
33. Lehmann, J.M.; Moore, L.B.; Smith-Oliver, T.A.; Wilkison, W.O.; Willson, T.M.; Kliewer, S.A. An Antidiabetic Thiazolidinedione Is a High Affinity Ligand for Peroxisome Proliferator-activated Receptor γ (PPARγ). *J. Biol. Chem.* **1995**, *270*, 12953–12956. [CrossRef] [PubMed]
34. Wang, D.; Haile, A.; Jones, L.C. Rosiglitazone-induced adipogenesis in a bone marrow mesenchymal stem cell line. In Proceedings of the 48th Annual Rocky Mountain Bioengineering Symposium and 48th International ISA Biomedical Sciences Instrumentation Symposium 2011, Denver, CO, USA, 15–17 April 2011; pp. 202–210.
35. Harmon, A.W.; Patel, Y.M.; Harp, J.B. Genistein inhibits CCAAT/enhancer-binding protein β (C/EBPβ) activity and 3T3-L1 adipogenesis by increasing C/EBP homologous protein expression. *Biochem. J.* **2002**, *367*, 203–208. [CrossRef] [PubMed]

36. Park, H.J.; Della-Fera, M.A.; Hausman, D.B.; Rayalam, S.; Ambati, S.; Baile, C.A. Genistein inhibits differentiation of primary human adipocytes. *J. Nutr. Biochem.* **2009**, *20*, 140–148. [CrossRef]
37. Chryssanthi, D.G.; Lamari, F.N.; Georgakopoulos, C.D.; Cordopatis, P. A new validated SPE-HPLC method for monitoring crocetin in human plasma–application after saffron tea consumption. *J. Pharm. Biomed. Anal.* **2011**, *55*, 563–568. [CrossRef]
38. Dludla, P.V.; Jack, B.; Viraragavan, A.; Pheiffer, C.; Johnson, R.; Louw, J.; Muller, C.J.F. A dose-dependent effect of dimethyl sulfoxide on lipid content, cell viability and oxidative stress in 3T3-L1 adipocytes. *Toxicol. Rep.* **2018**, *5*, 1014–1020. [CrossRef]
39. Kraus, N.A.; Ehebauer, F.; Zapp, B.; Rudolphi, B.; Kraus, B.J.; Kraus, D. Quantitative assessment of adipocyte differentiation in cell culture. *Adipocyte* **2016**, *5*, 351–358. [CrossRef]
40. Carmichael, J.; DeGraff, W.G.; Gazdar, A.F.; Minna, J.D.; Mitchell, J.B. Evaluation of a tetrazolium-based semiautomated colorimetric assay: Assessment of chemosensitivity testing. *Cancer Res* **1987**, *47*, 936–942.
41. Strober, W. Trypan Blue Exclusion Test of Cell Viability. *Curr. Protoc. Immunol.* **2015**, *111*, A3.B.1–A3.B.3. [CrossRef]
42. Bernlohr, D.A.; Bolanowski, M.A.; Kelly, T.J., Jr.; Lane, M.D. Evidence for an increase in transcription of specific mRNAs during differentiation of 3T3-L1 preadipocytes. *J. Biol. Chem.* **1985**, *260*, 5563–5567.
43. Feng, S.; Reuss, L.; Wang, Y. Potential of Natural Products in the Inhibition of Adipogenesis through Regulation of PPARγ Expression and/or Its Transcriptional Activity. *Molecules* **2016**, *21*, 1278. [CrossRef]
44. Cao, Z.; Umek, R.M.; McKnight, S.L. Regulated expression of three C/EBP isoforms during adipose conversion of 3T3-L1 cells. *Genes Dev.* **1991**, *5*, 1538–1552. [CrossRef] [PubMed]
45. Zhang, J.W.; Tang, Q.Q.; Vinson, C.; Lane, M.D. Dominant-negative C/EBP disrupts mitotic clonal expansion and differentiation of 3T3-L1 preadipocytes. *Proc. Natl. Acad. Sci. USA* **2004**, *101*, 43–47. [CrossRef] [PubMed]
46. Bradley, M.N.; Zhou, L.; Smale, S.T. C/EBPbeta regulation in lipopolysaccharide-stimulated macrophages. *Mol. Cell Biol.* **2003**, *23*, 4841–4858. [CrossRef] [PubMed]
47. Inoue, M.; Tanabe, H.; Matsumoto, A.; Takagi, M.; Umegaki, K.; Amagaya, S.; Takahashi, J. Astaxanthin functions differently as a selective peroxisome proliferator-activated receptor γ modulator in adipocytes and macrophages. *Biochem. Pharmacol.* **2012**, *84*, 692–700. [CrossRef]
48. Shirakura, Y.; Takayanagi, K.; Mukai, K.; Tanabe, H.; Inoue, M. β-Cryptoxanthin Suppresses the Adipogenesis of 3T3-L1 Cells via RAR Activation. *J. Nutr. Sci. Vitaminol.* **2011**, *57*, 426–431. [CrossRef]
49. Carmona, M.; Zalacain, A.; Sánchez, A.M.; Novella, J.L.; Alonso, G.L. Crocetin esters, picrocrocin and its related compounds present in Crocus sativus stigmas and Gardenia jasminoides fruits. Tentative identification of seven new compounds by LC-ESI-MS. *J. Agric. Food Chem.* **2006**, *54*, 973–979. [CrossRef]
50. Lobo, G.P.; Amengual, J.; Li, H.N.M.; Golczak, M.; Bonet, M.L.; Palczewski, K.; von Lintig, J. Beta,beta-carotene decreases peroxisome proliferator receptor gamma activity and reduces lipid storage capacity of adipocytes in a beta,beta-carotene oxygenase 1-dependent manner. *J. Biol. Chem.* **2010**, *285*, 27891–27899. [CrossRef]
51. Gul, T.; Balkhi, H.M.; Haq, E. Inhibition of Adipocyte Differentiation by Crocin in vitro Model of Obesity. *Saudi J. Life Sci.* **2017**, *2*, 306–311.
52. Rosen, E.D.; Walkey, C.J.; Puigserver, P.; Spiegelman, B.M. Transcriptional regulation of adipogenesis. *Genes Dev.* **2000**, *14*, 1293–1307.
53. Yamauchi, T.; Kamon, J.; Waki, H.; Murakami, K.; Motojima, K.; Komeda, K.; Ide, T.; Kubota, N.; Terauchi, Y.; Tobe, K.; et al. The mechanisms by which both heterozygous peroxisome proliferator-activated receptor gamma (PPARgamma) deficiency and PPARgamma agonist improve insulin resistance. *J. Biol. Chem.* **2001**, *276*, 41245–41254. [CrossRef] [PubMed]
54. Wang, N.D.; Finegold, M.J.; Bradley, A.; Ou, C.N.; Abdelsayed, S.V.; Wilde, M.D.; Taylor, L.R.; Wilson, D.R.; Darlington, G.J. Impaired energy homeostasis in C/EBP alpha knockout mice. *Science* **1995**, *269*, 1108–1112. [CrossRef] [PubMed]
55. Torres-Villarreal, D.; Camacho, A.; Castro, H.; Ortiz-Lopez, R.; de la Garza, A.L. Anti-obesity effects of kaempferol by inhibiting adipogenesis and increasing lipolysis in 3T3-L1 cells. *Am. J. Physiol. Biochem. Pharmacol.* **2019**, *75*, 83–88. [CrossRef] [PubMed]

56. Cooke, P.S.; Naaz, A. Effects of estrogens and the phytoestrogen genistein on adipogenesis and lipogenesis in males and females. *Birth Defects Res. Part A Clin. Mol. Teratol.* **2005**, *73*, 472–473. [CrossRef] [PubMed]
57. Zhang, M.; Ikeda, K.; Xu, J.-W.; Yamori, Y.; Gao, X.-M.; Zhang, B.-L. Genistein suppresses adipogenesis of 3T3-L1 cells via multiple signal pathways. *Phytother. Res.* **2009**, *23*, 713–718. [CrossRef]

foods

MDPI

Article

From By-Product to the Food Chain: Melon (*Cucumis melo* L.) Seeds as Potential Source for Oils

Adrián Rabadán [1], M. Antónia Nunes [2], Silvia M. F. Bessada [2], José E. Pardo [1], M. Beatriz P. P. Oliveira [2] and Manuel Álvarez-Ortí [1,*]

1 Higher Technical School of Agricultural and Forestry Engineering, University of Castilla-La Mancha, Campus Universitario s/n, 02071 Albacete, Spain; adrian.rabadan@uclm.es (A.R.); jose.pgonzalez@uclm.es (J.E.P.)

2 REQUIMTE/LAQV, Department of Chemical Sciences, Faculty of Pharmacy, University of Porto, R. Jorge Viterbo Ferreira, 228, 4050-313 Porto, Portugal; antonianunes.maria@gmail.com (M.A.N.); silviabessada@gmail.com (S.M.F.B.); beatoliv@ff.up.pt (M.B.P.P.O.)

* Correspondence: manuel.alvarez@uclm.es

Received: 19 August 2020; Accepted: 17 September 2020; Published: 23 September 2020

Abstract: Fruit-processing industries annually discard large volumes of fruit by-products. Thousands of tons of melon seeds could be recovered through the year from melon production. These seeds are an excellent source of vegetable oil with significant health-promoting properties due to their unsaturated fatty acid profile and high content of specific bioactive compounds. However, little information exists about the influence of melon cultivars and oil-extraction methods on oil characteristics. In this study, oils from nine different melon cultivars were evaluated. Additionally, two oil-extraction methods (screw and hydraulic press) were studied. Results showed that melon seeds may be used as a novel source of healthy oils. Higher-quality oils were obtained with the hydraulic press; however, low yields reduced industrial interest in this method. Oils extracted from the different cultivars showed high variability in the content of linoleic (51–69%) and oleic (15–34%) acids. Regarding vitamin E, γ-tocopherol was the main isoform found in melon-seed oils (99.81–456.73 mg/kg), followed by α- and δ-tocopherols. Significant concentrations of tocotrienols (α, β, and γ) were also found. Although all cultivars showed positive attributes, principal-component analysis (PCA) showed that Honey Dew and Blanco de Ribatejo could be specifically considered as a potential source of polyunsaturated oils with high concentrations of vitamin E.

Keywords: melon-seed oil; fruit waste; tocopherols; tocotrienols; unsaturated fatty acids; screw press

1. Introduction

The melon fruit (*Cucumis melo* L.) belongs to the *Cucurbitaceae* family, and it is grown in tropical and subtropical regions of the world. Global melon production has continuously risen in the last decade, reaching the current annual production of about 31.2×10^6 tons. Melon processing in the industry generates large quantities of by-products that are usually discarded. Within those by-products, melon seeds account for 10% of total melon weight [1]. However, melon seeds are not considered as waste in all regions of the world. In some Arabian countries, they are roasted and directly consumed [2], and in India they are dried and used to add flavor to traditional dishes and desserts [3]. This traditional use of seeds is not applied to melon production in Europe, where melon seeds are rarely used in the food chain.

Previous studies carried out mainly on melons grown in some developing countries confirmed the interest in melon seeds as a possible functional ingredient [3–5]. In this regard, the nutritional composition of melon-seed cultivars grown in different countries, including Egypt [6], Brazil [7–9], Tunisia [10], and China [11], was studied. With regard to its nutritional content, melon seeds were found to be a rich source of proteins (14.9–27.4%), lipids (25.7–30.8%), fiber (19.0–25.3%), carbohydrates (20.8–24.8%), and ashes (3.2–4.8%) [5,8,10]. Within proteins, melon seeds contain essential amino acids such as phenylalanine, isoleucine, and leucine [7,8]. However, results show important differences in the proximate and chemical composition of seeds depending on the studied melon cultivar. This is the case of Chinese hybrid ChunLi, which shows protein percentages of up to 29.9% and small concentrations of carbohydrates (5.6%) [11].

The oil content of melon seeds and the current demand for new vegetable oils have led to an excellent opportunity for the industrial production of vegetable oil from melon seeds. The selection of an appropriate extraction method is crucial for producing high-quality oils. Most previous research studied melon-seed oils extracted using solvents [5,7,9–12]. However, the use of solvents, mainly hexane, reduces oil quality and avoid their classification as virgin oils. On the other hand, the most modern method for seed-oil extraction consists of the use of supercritical fluids [3], but the production costs of this method are high, reducing its viability for industrial production. Within this framework, cold extraction based on the use of mechanical presses for oil extraction from seeds and nuts has resulted in the production of high-quality oils at affordable prices, encouraging its use for commercial purposes [13–16].

Oils from melon seeds are mainly composed of unsaturated fatty acids, where linoleic and oleic acids are predominant. Most studies reported a content of linoleic acid ranging from 64.1% to 69.0%, and oleic acid from 13.7% to 19.4% [8,10,17]. However, these percentages may differ in some cultivars whit lower levels in linoleic acid and higher in oleic acid, like those reported by da Silva and Jorge [9], where 59.0% of linoleic acid and 26.4% of oleic acid were found in an undetermined cultivar, or those obtained by De Mello, Bora, and Narain [7], where similar percentages were found in the Daimiel cultivar (cv.). Regarding saturated fatty acids, melon-seed oil generally shows low percentages, ranging from 8.7% to 10.2% [7–10,17], although in some cultivars like ChunLi, a concentration of palmitic acid of up to 23.9% was reported [11].

Beyond the well-known beneficial effects of polyunsaturated oils for human health [18], analysis of bioactive compounds in oils is crucial. Within minor components, significant amounts of tocopherols were reported in melon-seed oils. Tocopherols and tocotrienols are part of vitamin E, a potent antioxidant that was reported to protect against cancer and bone, cardiovascular, eye, nephrological, and neurological diseases [19]. In previous studies, the average total tocopherol content in melon-seed oils was variable, with amounts between 270 and 720 mg/kg [9,10,20]. γ-tocopherol was the main tocopherol isoform described in melon oils, while δ-tocopherol was reported in the Canary melon and in cv. Maazoun, but not in other cultivars [9]. For α-tocopherol, which is the most active homologous in humans, minor concentrations (22.0–68.8 mg/kg) were reported [9,10]. With regard to tocotrienols, little information exists, as they account for roughly 1% of the total studies on vitamin E [21]. The only information about tocotrienols in melon oil was provided by Górnaś, Soliven, and Segliņa [20], who reported a total concentration of 13.7 mg/kg, with the dominance of γ-tocotrienol.

Considering the high variability of the results described in the different analyzed cultivars, further analysis of the main melon cultivars grown in Europe is needed, encouraging the return of these agroindustrial residues into the food chain. In this study, nine different *C. melo* cultivars, including three types of traditional Spanish cultivar Piel de Sapo, are evaluated. Furthermore, two pressure-extraction methods were used, and the obtained results were compared regarding their availability for the industrial extraction of high-quality oils.

2. Materials and Methods

2.1. Plant Material

Nine different melon cultivars were analyzed: Amarillo Oro Canario, Arizo, Blanco de Ribatejo, Charentais, Honeydew, Piñonet, Tendral Valenciano, Tendral Verde Tardío, and Piel de Sapo. Seeds from cv. Piel de Sapo were obtained from three different conditions: traditional cultivation, organic cultivation, and seeds from Protected Geographical Indication (PGI) Melon de la Mancha, kindly supplied by the Regulatory Board of the PGI (Tomelloso, Spain). Commercial melon seeds were obtained from local suppliers in Albacete (Spain). Seeds were cleaned and washed to remove sugars and any adhered residues. Then, seeds were dried at room temperature for several days until the seeds from all cultivars reached a moisture of less than 10%.

To evaluate the proportion of the peel with respect to the total weight of the seeds, 100 seeds were selected from each cultivar and manually peeled. The peels were weighted, and the proportion of the peel was calculated as the peel weight divided by the total seed weight.

2.2. Oil Extraction

Oil extraction was carried out by using a Komet Oil Press CA59G screw press (IBG Monforts Oekotec GmbH & Co. KG, Monchengladbach, Germany). One kilogram of unpeeled seeds was introduced directly into the press once the barrel was heated to 100 °C to ensure the correct extraction of oil [14]. Medium rotational-speed conditions were selected (49 rpm). Additionally, seeds from cv. Piel de Sapo PGI were subjected to extraction by using a hydraulic press (MECAMAQ model DEVF 80, Vila-Sana, Lleida, Spain). For extraction with the hydraulic press, 1 kg of ground unpeeled seeds was placed on the press, and the seeds were subjected to a pressure of 150 bar for 10 min. After pressing, oil was centrifuged to remove remaining solids. Oil samples were stored in dark glass bottles at 5 °C to avoid degradation until analysis.

2.3. Regulated Quality Parameters

Regulated quality parameters consist of free acidity and peroxide values. To determine free acidity, expressed as % of oleic acid, a solution of melon-seed oil dissolved in ethanol/ether (1:1) was titrated with a 0.1 mol/L potassium hydroxide ethanolic solution [22]. On the other hand, the peroxide value, expressed in milliequivalents of active oxygen per kilogram of oil (meq O_2/kg), was measured according to European Union (EU) regulations [22]. Briefly, chloroform and acetic acid were added to an oil sample, mixed vigorously, and left to react with a solution of potassium iodide in the darkness. Then, the free iodine was titrated with a sodium thiosulfate solution [22].

2.4. Fatty Acid Profile

Fatty acid profile was measured according to Santos et al. [23]. Briefly, 2 mL of n-hexane was added to 0.02 g of oil to obtain fatty acid methyl esters (FAME) by cold transmethylation with methanolic potassium hydroxide. Then, 200 μL of methanolic potassium hydroxide solution (2 N) was added and vigorously mixed. Then, the supernatant was carefully transferred to a glass vial and analyzed by gas chromatography in a Shimadzu GC-2010 Plus Gas Chromatograph (Shimadzu, Tokyo, Japan). This was performed using a CPSil 88 fused silica capillary column (50 m × 0.25 mm i.d.), 0.20 μm film thickness (Varian, Middelburg, The Netherlands), and helium was used as the carrier gas (120 kPa). The used temperature program was a first step of 5 min at 140 °C, followed by an increase of 5 °C/min from 140 to 220 °C, and then maintaining at 220 °C for 15 min. The temperature of the injector and detector was 250 and 270 °C, respectively, and the split ratio was 1:50 with an injection volume of 1 μL. Lastly, each FAME was identified by direct comparison with a standard mixture (FAME 37, Supelco, Bellefonte, PA, USA). All analyses were performed in duplicate, and results are expressed as the relative percentage of each FA on the basis of relative peak areas.

2.5. Vitamin E Determination

Analogously, vitamin E values were determined by HPLC analysis in oil samples according to Alves et al. [24]. Briefly, about 20 mg of oil was diluted in 1 mL of n-hexane (HPLC-grade, Merck, Darmstadt, Germany), where 20 μg/mL of tocol was added as internal standard. Then, 20 μL was injected to perform the separation on a normal-phase SupelcosilTM LC-SI column (3 μm; 75 × 3.0 mm; Supelco, Bellefonte, PA, USA). The used equipment was an HPLC system (Jasco, Tokyo, Japan) equipped with an AS-2057 automated injector, a PU-2089 pump, and an MD-2018 multiwavelength diode array detector (DAD) coupled with an FP-2020 fluorescence detector (Jasco, Japan). They were programmed for excitation at 290 nm and emission at 330 nm. Lastly, the identification of the compounds was accomplished by a comparison with commercial standards. Analyses were performed in duplicate, and results are expressed as mg/kg of oil.

2.6. Color Determination

Oil samples were filtered, and color was measured using a UV/Vis Jasco V-530 spectrophotometer (Jasco Analytical, Madrid, Spain). Basically, oil samples were placed in quartz cuvettes (1 cm path length) for analysis, using N-hexane as the blank reference. The obtained values were used to calculate CIELAB chromatic coordinates: L* (brightness), a* (red–green component), b* (yellow–blue component) as recommended by the Commission Internationale de l'Eclairage (CIE, Wien, Austria) [25].

2.7. Statistical Analysis

Data are expressed as mean ± standard deviation of the obtained results for the selected cultivars. Data were analyzed using the *t*-test and Duncan's test. Statistical significance was defined for $p < 0.05$ (95% confidence level). Pearson's correlations were examined. To perform principal-component analysis (PCA) for melon-seed cultivars, those variables that had previously shown significant differences were used. The Kaiser–Meyer–Olkin (KMO) test for sampling adequacy was used. All statistical analyses were carried out using the SPSS program v. 23.0 for Windows.

3. Results and Discussion

3.1. Oil-Extraction Yield

First, for a clearer idea about oil yield, it is important to consider seed parts that do not contain a significant composition in oil, which may contribute to reducing the yield value. In this sense, seed peel is mainly composed of carbohydrates, especially fiber [10]. The melon-seed peel constitutes about 30–40/100 g of seed weight. Significant differences were found between cultivars (Table 1). Cv. Piel de Sapo grown under conventional production (not organic) showed the lowest proportion of seed peel (30.65/100 g) followed by the organic Piel de Sapo. Cv. Piel de Sapo grown under PGI conditions showed one of the highest proportions of seed peel (39.00/100 g).

Table 1. Proportion of seed peels and oil-extraction yields obtained with screw press in selected melon cultivars. PGI, Protected Geographical Indication.

Cultivar	Peel (g/100 g seeds)	Oil-Extraction Yield (g/100 g seeds)
Amarillo Oro Canario	36.02 ± 0.47 ^{e}	21.43 ± 1.65 ^{de}
Arizo	39.51 ± 0.33 ^{ab}	16.95 ± 1.26 ^{g}
Blanco de Ribatejo	37.41 ± 0.49 ^{d}	19.64 ± 0.77 ^{ef}
Charentais	32.79 ± 1.19 ^{f}	21.43 ± 1.21 ^{de}
Honeydew	38.15 ± 0.12 ^{cd}	21.67 ± 0.79 ^{d}
Piñonet	33.00 ± 0.21 ^{f}	29.90 ± 0.54 ^{a}
Tendral Valenciano	38.45 ± 0.34 ^{bcd}	19.36 ± 1.72 ^{f}
Tendral Verde Tardío	39.85 ± 0.23 ^{a}	18.92 ± 0.84 ^{f}
Piel de Sapo conventional	30.65 ± 0.10 ^{g}	24.59 ± 0.73 ^{bc}
Piel de Sapo organic	32.89 ± 0.20 ^{f}	24.08 ± 0.81 ^{c}
Piel de Sapo PGI	39.00 ± 1.36 ^{abc}	26.23 ± 1.08 ^{b}

Mean ± standard deviation; $^{a–g}$ different letters in same column represent significant differences, $p < 0.05$ between samples.

Melon seeds were identified as a good source of oil, with percentages of lipids about 30.7%–32.3% [8–10]. Oil yields from melon seeds were high enough to encourage their use for oil-production purposes when a screw press is used (Table 1). Obtained yields using the screw press were statistically different depending on the considered cultivar. Cvs. Piñonet and Piel de Sapo PGI showed the highest values, 29.90 and 26.23 g per 100 g, respectively. The lowest values were reported in the Arizo cultivar (16.95/100 g seeds). Negative but not statistically significant ($r = -0.576$, $p = 0.64$) correlation was found between peel proportion and oil yield obtained with the screw press.

3.2. Oil-Extraction Methods, Oil Quality, and Color

To evaluate the differences regarding extraction systems, seeds from cv. Piel de Sapo PGI were subjected to oil extraction with two presses, a hydraulic and a screw press. Extraction with the hydraulic press was performed under room temperature, while extraction with the screw press requires previous heating to obtain optimal performance. The data regarding oil yield, regulated quality (acidity and peroxide index) and color of the oils obtained are shown in Table 2. The pressure system selected for oil extraction had significant influence on oil quality and yield. Oil extraction in cv. Piel de Sapo PGI using the screw press resulted in an oil yield of 26.23/100 g, while extraction with the hydraulic press was almost four times smaller (6.80/100 g). This low yield makes the hydraulic press unsuitable for obtaining an economic benefit. Therefore, for analysis of the profile of fatty acids and vitamin E for the rest of the cultivars, only oils extracted with the screw press were used, since this could be the most appropriate method for obtaining commercial oils in the industry.

Table 2. Oil yield, parameters of regulated quality, and color of melon-seed oils (cv. Piel de Sapo PGI) according to extraction method.

Parameters	Hydraulic Press	Screw Press
Oil yield	6.80 ± 0.63 [b]	26.23 ± 1.08 [a]
Regulated quality	-	-
Acidity	0.30 ± 0.04 [b]	0.41 ± 0.05 [a]
Peroxide Index	0.00	0.00
Color	-	-
L*	91.28 ± 0.59 [b]	92.75 ± 0.60 [a]
a*	-4.70 ± 0.41 [a]	-7.40 ± 0.90 [b]
b*	20.97 ± 1.44 [b]	29.62 ± 1.46 [a]

Mean ± standard deviation; [a,b] different letters in the same line represent significant differences $p < 0.05$ between samples.

Regarding oil quality, Codex Alimentarius [26] does not have specific regulation for melon-oil quality standards. Results showed that oils obtained with the hydraulic press were of slightly better quality than that of oils obtained with the screw press. In all cases, the values in oils obtained using pressure systems were significantly lower than the values reported for melon-seed oils obtained using solvent extraction [7,11].

Melon oils show a light yellow color. Nevertheless, CIELAB color parameters showed differences attending to the used extraction method. Oils obtained with the screw press showed more intense yellow colors, with higher values for the b* parameter (Table 2). Oil extraction using the screw press requires high temperatures applied on the barrel to ensure proper oil extraction [27]. The processing temperature of screw press compared to room temperature used in hydraulic extraction may affect oil pigment content [28]. Previous studies on plant oils showed that lutein, which provides the yellow color to oils, is more resistant to high temperatures than other pigments are, such as chlorophylls [29]. Furthermore, some studies even reported an increase in the content of lutein after the thermal processing of food products due to the inactivation of enzymes responsible for oxidizing carotenoids [30]. Although the total content of carotenoids in melon oil was reported to be low [9], the balance of carotenoids in melon oil could be the reason for the observed change of color in melon oils depending on the extraction method.

3.3. Fatty Acids

As previously reported, melon-seed oils are mainly composed of linoleic (50.67%–69.22%) and oleic (15.23%–33.96%) acids. Saturated fatty acids, mainly palmitic and stearic, accounted for less than 15.62% in all studied cultivars (Table 3). Our data support previous results about the high variability of the fatty acid profile in melon-seed oils [5,7,8,10,11]. Some cultivars, such as Tendral Valenciano and Tendral Verde, showed a high content of linoleic acid, 69.22% and 69.15%, respectively, in comparison with cv. Piel de Sapo PGI, which showed the lowest values. As widely reported in other unsaturated plant oils, linoleic and oleic content were negatively correlated [31,32].

Extraction method had no effect on the fatty acid composition of the oils. When seeds from cv. Piel de Sapo PGI were used, oils extracted with the screw and hydraulic presses showed slight differences in fatty acid profile (Table 3). Similarly, small differences in the content of linoleic acid were found when the results for cv. Honey Dew were compared to those of the study of Bora, Narain, and de Melio [5], who used solvent extraction for the same cultivar. In all cases, the differences reported from the cultivars were determinants compared to those small differences that could be attributed to the oil-extraction method.

3.4. Vitamin E

Regarding vitamin E content, the studied melon-seed oils showed significant differences (Table 4). Cvs. Honey Dew and Blanco de Ribatejo showed the highest contents of vitamin E, with 530.62 and 468.19 mg/kg, respectively. γ- and α-tocopherols were the main components of vitamin E. The amounts of α-tocopherol (37.42–74.71 mg/kg) was significantly higher than those in some previous studies for specific cultivars [9,10], but were in accordance with the data reported by Górnaś and Rudzińska [12]. α-tocopherol is crucial for oil quality, as it is the form preferentially absorbed and accumulated in humans. Cv. Tendral valenciano, which showed a small concentration of vitamin E, was, however, the one with the highest content in α-tocopherol. As reported with regard to oils from dicotyledonous plants, tocotrienol content was low compared to tocopherol content [33]; however, it was higher than that reported in some oils with similar fatty acid profiles, such as walnut oil [34,35]. γ-tocotrienol was the main tocotrienol found, in agreement with the study of Górnaś, Soliven, and Segliņa [20]. The total concentration of vitamin E in melon oils and the content of the specific isoforms could be used for the authentication of products containing melon oils and flours as functional ingredients.

Oil-seed extraction methods significantly affect the concentration of tocopherols in oils, especially if solvent extraction is compared to pressing [36–38]. Our results showed that significant differences appear in the concentration of tocopherol and tocotrienol forms, and in the total content of vitamin E. The vitamin E content of the Piel de Sapo PGI cultivar, extracted by hydraulic pressing, was lower than the content obtained with screw pressing; however, as reported for fatty acids, cultivar had a larger effect on vitamin E content than extraction method did.

3.5. Principal-Component Analysis

The reported variability in the oil composition of cultivars is reflected in principal-component analysis (Figure 1). Principal Component 1 (PC1) was mainly composed of fatty acids (C18:2n6c; C18:1n9c; C18:0) and tocotrienols (β-T3), while Principal Component 2 (CP2) was composed of tocopherols (α-T; γ-T). The negative correlation between linoleic and oleic acid was clear. Similar negative correlation could be observed for the α- and γ-tocopherols. As γ-tocopherol is the main component of vitamin E in melon-seed oils, and major interest exists for oils with a high content in unsaturated fatty acids, the cultivars in the top right of Figure 1 are the most interesting for melon-seed-oil production. These are cvs. Charentais and Honey Dew. However, the provided information also encourages the production of oils with more monounsaturated fatty acids and with a higher concentration of the most active homologous α-tocopherol. In this case, the use of cvs. Piñonet and Piel de Sapo PGI, in the bottom left of Figure 1, is preferable.

Table 3. Fatty acid profile (g/100 g of oil) of obtained melon-seed oils. All oil samples were extracted with screw press except the last column, which shows results of fatty acid profile of oil from cv Piel de Sapo (PGI), extracted with hydraulic press.

	Amarillo Oro Canario	Arizo	Blanco de Ribatejo	Charentais	Honey Dew	Piñonet	Tendral Valenciano	Tendral Verde Tardío	Piel de Sapo Conventinal	Piel de Sapo Organic	Piel de Sapo PGI	Piel de Sapo PGI Hydraulic
C14:0	0.04 ± 0.00 ^{f}	0.05 ± 0.00 ^{c}	0.06 ± 0.00 ^{b}	0.04 ± 0.00 ^{g}	0.05 ± 0.00 ^{d}	0.06 ± 0.00 ^{a}	0.04 ± 0.00 ^{e}	0.05 ± 0.00 ^{c}	0.06 ± 0.00 ^{a}	0.04 ± 0.00 ^{ef}	0.04 ± 0.00 ^{h}	0.04 ± 0.00 ^{h}
C15:0	0.03 ± 0.00 ^{f}	0.03 ± 0.00 ^{ab}	0.03 ± 0.00 ^{a}	0.03 ± 0.00 ^{cd}	0.02 ± 0.00 ^{f}	0.03 ± 0.00 ^{bc}	0.03 ± 0.00 ^{bc}	0.03 ± 0.00 ^{d}	0.03 ± 0.00 ^{cd}	0.03 ± 0.00 ^{c}	0.02 ± 0.00 ^{e}	0.03 ± 0.00 ^{e}
C16:0	8.59 ± 0.01 ^{h}	9.39 ± 0.02 ^{e}	9.43 ± 0.00 ^{c}	9.74 ± 0.00 ^{a}	7.19 ± 0.01 ^{k}	8.22 ± 0.01 ^{j}	9.40 ± 0.02 ^{d}	9.72 ± 0.00 ^{b}	8.65 ± 0.01 ^{g}	8.25 ± 0.00 ^{i}	9.10 ± 0.01 ^{f}	9.10 ± 0.01 ^{f}
C16:1	0.09 ± 0.00 ^{e}	0.08 ± 0.00 ^{i}	0.09 ± 0.00 ^{h}	0.10 ± 0.00 ^{d}	0.06 ± 0.00 ^{k}	0.08 ± 0.00 ^{h}	0.10 ± 0.00 ^{b}	0.10 ± 0.00 ^{c}	0.08 ± 0.00 ^{h}	0.08 ± 0.00 ^{j}	0.12 ± 0.00 ^{a}	0.12 ± 0.00 ^{a}
C17:0	0.07 ± 0.00 ^{cd}	0.08 ± 0.00 ^{a}	0.08 ± 0.00 ^{ab}	0.07 ± 0.00 ^{de}	0.06 ± 0.00 ^{e}	0.07 ± 0.00 ^{bcd}	0.07 ± 0.00 ^{cd}	0.07 ± 0.00 ^{abc}	0.07 ± 0.00 ^{de}	0.08 ± 0.00 ^{a}	0.08 ± 0.00 ^{ab}	0.07 ± 0.00 ^{abc}
C17:1	0.03 ± 0.00 ^{a}	0.03 ± 0.00 ^{a}	0.01 ± 0.00 ^{e}	0.01 ± 0.00 ^{cd}	0.01 ± 0.00 ^{g}	0.01 ± 0.00 ^{fg}	0.02 ± 0.00 ^{b}	0.01 ± 0.00 ^{c}	0.01 ± 0.00 ^{cde}	0.01 ± 0.00 ^{f}	0.01 ± 0.00 ^{cd}	0.01 ± 0.00 ^{de}
C18:0	5.15 ± 0.01 ^{i}	5.66 ± 0.01 ^{c}	5.47 ± 0.01 ^{e}	5.29 ± 0.01 ^{g}	5.65 ± 0.00 ^{d}	5.67 ± 0.00 ^{b}	4.57 ± 0.01 ^{k}	5.01 ± 0.00 ^{j}	5.21 ± 0.00 ^{h}	5.86 ± 0.01 ^{a}	5.35 ± 0.00 ^{f}	5.36 ± 0.00 ^{f}
C18:1n9c	18.66 ± 0.03 ^{i}	27.85 ± 0.03 ^{d}	21.06 ± 0.00 ^{g}	15.60 ± 0.02 ^{k}	22.05 ± 0.03 ^{f}	31.65 ± 0.01 ^{c}	15.98 ± 0.00 ^{j}	15.23 ± 0.00 ^{l}	18.88 ± 0.00 ^{h}	25.83 ± 0.04 ^{e}	33.96 ± 0.03 ^{a}	33.78 ± 0.04 ^{b}
C18:2n6c	66.83 ± 0.05 ^{d}	56.02 ± 0.01 ^{i}	63.11 ± 0.01 ^{g}	68.44 ± 0.00 ^{c}	64.31 ± 0.01 ^{f}	53.59 ± 0.00 ^{j}	69.22 ± 0.04 ^{a}	69.15 ± 0.00 ^{b}	66.37 ± 0.00 ^{e}	59.13 ± 0.05 ^{h}	50.69 ± 0.03 ^{l}	50.87 ± 0.03 ^{k}
C20:0	0.17 ± 0.00 ^{h}	0.28 ± 0.00 ^{a}	0.22 ± 0.00 ^{e}	0.24 ± 0.00 ^{c}	0.18 ± 0.00 ^{g}	0.23 ± 0.00 ^{c}	0.18 ± 0.00 ^{g}	0.21 ± 0.00 ^{f}	0.22 ± 0.00 ^{e}	0.25 ± 0.00 ^{b}	0.23 ± 0.00 ^{d}	0.23 ± 0.00 ^{d}
C18:3n3	0.16 ± 0.00 ^{i}	0.26 ± 0.00 ^{a}	0.22 ± 0.00 ^{c}	0.23 ± 0.00 ^{b}	0.17 ± 0.00 ^{g}	0.14 ± 0.00 ^{k}	0.19 ± 0.00 ^{f}	0.21 ± 0.00 ^{d}	0.20 ± 0.00 ^{e}	0.20 ± 0.00 ^{e}	0.16 ± 0.00 ^{h}	0.15 ± 0.00 ^{j}
C20:1n9	0.12 ± 0.00 ^{f}	0.13 ± 0.00 ^{d}	0.12 ± 0.00 ^{e}	0.11 ± 0.00 ^{i}	0.16 ± 0.00 ^{a}	0.15 ± 0.00 ^{b}	0.11 ± 0.00 ^{j}	0.11 ± 0.00 ^{h}	0.12 ± 0.00 ^{g}	0.14 ± 0.00 ^{c}	0.13 ± 0.00 ^{d}	0.13 ± 0.00 ^{d}
C21:0	0.00 ± 0.00 ^{d}	0.01 ± 0.00 ^{ab}	0.01 ± 0.00 ^{a}	0.01 ± 0.00 ^{a}	0.00 ± 0.00 ^{d}	0.00 ± 0.00 ^{c}	0.00 ± 0.00 ^{e}	0.01 ± 0.00 ^{ab}	0.01 ± 0.00 ^{ab}	0.00 ± 0.00 ^{c}	0.01 ± 0.00 ^{b}	0.01 ± 0.00 ^{b}
C20:2	0.01 ± 0.00 ^{d}	0.01 ± 0.00 ^{e}	0.01 ± 0.00 ^{cd}	0.01 ± 0.00 ^{bc}	0.02 ± 0.00 ^{a}	0.01 ± 0.00 ^{e}	0.01 ± 0.00 ^{b}	0.01 ± 0.00 ^{b}	0.01 ± 0.00 ^{bc}	0.01 ± 0.00 ^{cd}	0.01 ± 0.00 ^{e}	0.01 ± 0.00 ^{e}
C22:0	0.03 ± 0.00 ^{i}	0.07 ± 0.00 ^{a}	0.05 ± 0.00 ^{d}	0.05 ± 0.00 ^{d}	0.04 ± 0.00 ^{h}	0.05 ± 0.00 ^{e}	0.04 ± 0.00 ^{g}	0.05 ± 0.00 ^{f}	0.05 ± 0.00 ^{d}	0.05 ± 0.00 ^{c}	0.06 ± 0.00 ^{b}	0.06 ± 0.00 ^{b}
C24:0	0.05 ± 0.00 ^{cd}	0.07 ± 0.01 ^{a}	0.05 ± 0.00 ^{b}	0.04 ± 0.00 ^{de}	0.05 ± 0.00 ^{bc}	0.04 ± 0.00 ^{ef}	0.04 ± 0.00 ^{g}	0.04 ± 0.00 ^{g}	0.04 ± 0.00 ^{fg}	0.05 ± 0.00 ^{bc}	0.04 ± 0.00 ^{de}	0.04 ± 0.00 ^{de}
SFA	14.11 ± 0.03 ^{i}	15.62 ± 0.01 ^{a}	15.39 ± 0.02 ^{c}	15.50 ± 0.02 ^{b}	13.23 ± 0.02 ^{j}	14.38 ± 0.01 ^{g}	14.37 ± 0.03 ^{g}	15.18 ± 0.01 ^{d}	14.33 ± 0.01 ^{h}	14.61 ± 0.01 ^{f}	14.92 ± 0.01 ^{e}	14.92 ± 0.01 ^{e}
MUFA	18.89 ± 0.03 ^{i}	28.09 ± 0.03 ^{d}	21.28 ± 0.00 ^{g}	15.82 ± 0.02 ^{k}	22.27 ± 0.03 ^{f}	31.88 ± 0.01 ^{c}	16.21 ± 0.01 ^{j}	15.45 ± 0.00 ^{l}	19.09 ± 0.00 ^{h}	26.04 ± 0.05 ^{e}	34.22 ± 0.03 ^{a}	34.04 ± 0.04 ^{b}
PUFA	67.00 ± 0.05 ^{d}	56.29 ± 0.02 ^{i}	63.34 ± 0.01 ^{g}	68.69 ± 0.01 ^{c}	64.50 ± 0.01 ^{f}	53.74 ± 0.01 ^{j}	69.43 ± 0.04 ^{a}	69.37 ± 0.01 ^{b}	66.58 ± 0.01 ^{e}	59.35 ± 0.05 ^{h}	50.86 ± 0.03 ^{l}	51.03 ± 0.03 ^{k}

Mean ± standard deviation; $^{a–l}$ different letters in same line represent significant differences; $p < 0.05$ between samples.

Table 4. Tocopherol and tocotrienol contents (milligrams per kilogram) of studied cultivars. All oils samples were extracted with screw press except the last row, which shows results of tocopherols and tocotrienols of oil from cv. Piel de Sapo (PGI) extracted with hydraulic press.

Cultivar	α-T	α-T3	γ-T	β-T3	γ-T3	δ-T	Total Vit E
Amarillo Oro Canario	56.04 ± 1.81 ^{e}	11.38 ± 0.42 ^{c}	278.26 ± 4.63 ^{d}	13.94 ± 1.12 ^{de}	11.31 ± 0.62 ^{d}	25.84 ± 0.88 ^{b}	396.76 ± 1.45 ^{d}
Arizo	37.42 ± 1.16 ^{h}	-	358.45 ± 0.07 ^{b}	18.90 ± 0.86 ^{b}	12.34 ± 0.02 ^{bc}	22.43 ± 0.08 ^{c}	449.55 ± 2.00 ^{c}
Blanco de Ribatejo	59.95 ± 1.50 ^{d}	12.62 ± 0.57 ^{ab}	341.15 ± 7.53 ^{c}	17.52 ± 0.63 ^{c}	13.64 ± 0.58 ^{a}	23.33 ± 0.88 ^{c}	468.19 ± 3.37 ^{b}
Charentais	38.61 ± 1.24 ^{h}	-	217.39 ± 1.44 ^{e}	21.26 ± 0.34 ^{a}	11.91 ± 0.65 ^{cd}	27.17 ± 0.58 ^{a}	316.33 ± 1.38 ^{e}
Honey Dew	41.63 ± 0.51 ^{g}	-	456.73 ± 1.30 ^{a}	18.93 ± 0.72 ^{b}	-	13.34 ± 0.11 ^{h}	530.62 ± 2.43 ^{a}
Piñonet	61.58 ± 0.21 ^{d}	12.63 ± 0.63 ^{ab}	128.28 ± 0.28 ^{h}	11.53 ± 0.28 ^{f}	-	17.82 ± 0.40 ^{e}	231.83 ± 1.24 ^{jk}
Tendral Valenciano	74.71 ± 3.05 ^{a}	12.87 ± 0.34 ^{a}	99.81 ± 1.43 ^{i}	17.20 ± 0.23 ^{c}	12.79 ± 0.16 ^{b}	17.80 ± 0.78 ^{e}	235.18 ± 6.00 ^{j}
Tendral Verde Tardío	55.11 ± 0.43 ^{e}	11.91 ± 0.18 ^{bc}	143.84 ± 1.10 ^{g}	20.79 ± 0.13 ^{a}	12.26 ± 0.53 ^{bc}	19.48 ± 0.51 ^{d}	263.40 ± 1.51 ^{h}
Piel de sapo conventional	59.91 ± 0.21 ^{d}	-	180.49 ± 4.35 ^{f}	20.37 ± 0.53 ^{a}	13.04 ± 0.53 ^{ab}	16.26 ± 0.29 ^{f}	290.06 ± 3.85 ^{g}
Piel de sapo organic	52.47 ± 0.99 ^{f}	-	212.81 ± 0.52 ^{e}	14.24 ± 0.04 ^{d}	12.99 ± 0.07 ^{ab}	14.78 ± 1.08 ^{g}	307.29 ± 0.39 ^{f}
Piel de sapo PGI	64.86 ± 0.30 ^{c}	12.60 ± 0.27 ^{ab}	128.44 ± 1.77 ^{h}	13.02 ± 0.10 ^{e}	13.69 ± 0.15 ^{a}	9.40 ± 0.22 ^{i}	241.99 ± 1.67 ^{i}
Piel de sapo PGI HP	68.86 ± 0.46 ^{b}	-	125.35 ± 2.03 ^{h}	13.45 ± 0.02 ^{de}	12.70 ± 0.48 ^{bc}	9.33 ± 0.79 ^{i}	229.70 ± 1.24 ^{k}

Mean ± standard deviation; $^{a-k}$ different letters in the same column represent significant differences; $p < 0.05$ between samples.

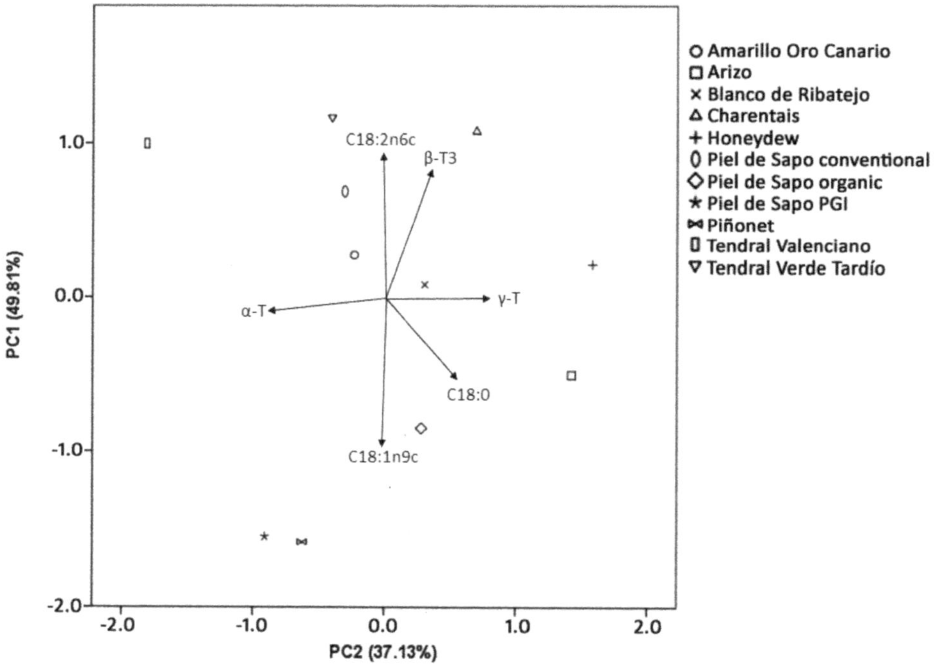

Figure 1. Principal-component analysis of oils from selected melon-seed cultivars.

4. Conclusions

Melon-seed oil is proposed as a highly valuable product that can be obtained from the byproducts of the agroindustrial processing of melons. This oil showed a high content of polyunsaturated fatty acids, mainly linoleic, and a high concentration of vitamin E. However, the effect of cultivar and oil-extraction method must be considered, as they have crucial influence in melon-oil characteristics. The oil-extraction method influences oil-quality parameters and oil due to the processing temperature in screw-press extraction. However, low yields obtained by hydraulic-press extraction could be inconvenient for industries to obtain an economic benefit. On the other hand, the use of different cultivars results in oils with different degrees of unsaturation and vitamin E content. Regarding vitamin E content, melon-seed oil may be considered as a rich source of tocopherols and tocotrienols.

The industrial extraction of oil from melon seeds is a feasible option to obtain high-quality oil in order to meet the current demand of vegetable oils for human nutrition.

Author Contributions: Conceptualization, A.R. and M.Á.-O.; methodology, M.A.N., M.B.P.P.O., and M.Á.-O.; formal analysis, M.A.N., S.M.F.B., and M.Á.-O.; investigation, A.R., M.A.N., and M.Á.-O.; resources, A.R., J.E.P., and M.B.P.P.O.; data curation, A.R., M.A.N., and S.M.F.B.; writing—original-draft preparation, A.R.; writing—review and editing, M.Á.-O.; visualization, J.E.P. and M.B.P.P.O.; supervision, J.E.P. and M.B.P.P.O.; project administration, J.E.P.; funding acquisition, A.R., J.E.P., and M.B.P.P.O. All authors have read and agreed to the published version of the manuscript.

Funding: This research was partially supported by a grant from the University of Castilla-La Mancha (ref. [2015/4062]) and by a project from the Castilla-La Mancha Regional Government (FEDER-SBPLY/19/180501/000047).

Conflicts of Interest: The authors declare no conflict of interest.

References

1. Rashid, U.; Rehman, H.A.; Hussain, I.; Ibrahim, M.; Haider, M.S. Muskmelon (*Cucumis melo*) seed oil: A potential non-food oil source for biodiesel production. *Energy* **2011**, *36*, 5632–5639. [CrossRef]
2. Al-Khalifa, A.S. Physicochemical characteristics, fatty acid composition, and lipoxygenase activity of crude pumpkin and melon seed oils. *J. Agric. Food Chem.* **1996**, *44*, 964–966. [CrossRef]
3. Maran, J.P.; Priya, B. Supercritical fluid extraction of oil from muskmelon (*Cucumis melo*) seeds. *J. Taiwan Inst. Chem. Eng.* **2015**, *47*, 71–78. [CrossRef]
4. Girgis, P.; Said, F. Lesser known Nigerian edible oils and fats. I.—Characteristics of melon seed oils. *J. Sci. Food Agric.* **1968**, *19*, 615–616. [CrossRef]
5. Bora, P.S.; Narain, N.; de Melio, M.L.S. Characterization of the seed oils of some commercial cultivars of melon. *Eur. J. Lipid Sci. Technol.* **2000**, *102*, 266–269. [CrossRef]
6. El-Magoli, S.B.; Morad, M.M.; El-Fara, A.A. Evaluation of Some Egyptian Sweet Melon Seed Oils. *Fette Seifen Anstrichm.* **1979**, *81*, 201–203. [CrossRef]
7. De Mello, M.L.S.; Bora, P.S.; Narain, N. Fatty and Amino Acids Composition of Melon (*Cucumis melo* Var. saccharinus) Seeds. *J. Food Compos. Anal.* **2001**, *14*, 69–74. [CrossRef]
8. De Melo, M.L.S.; Narain, N.; Bora, P.S. Characterisation of some nutritional constituents of melon (*Cucumis melo* hybrid AF-522) seeds. *Food Chem.* **2000**, *68*, 411–414. [CrossRef]
9. da Silva, A.C.; Jorge, N. Bioactive compounds of oils extracted from fruits seeds obtained from agroindustrial waste. *Eur. J. Lipid Sci. Technol.* **2017**, *119*. [CrossRef]
10. Mallek-Ayadi, S.; Bahloul, N.; Kechaou, N. Chemical composition and bioactive compounds of *Cucumis melo* L. seeds: Potential source for new trends of plant oils. *Process Saf. Environ. Prot.* **2018**, *113*, 68–77. [CrossRef]
11. Mian-Hao, H.; Yansong, A. Characteristics of some nutritional composition of melon (*Cucumis melo* hybrid 'ChunLi') seeds. *Int. J. Food Sci. Technol.* **2007**, *42*, 1397–1401. [CrossRef]
12. Górnaś, P.; Rudzińska, M. Seeds recovered from industry by-products of nine fruit species with a high potential utility as a source of unconventional oil for biodiesel and cosmetic and pharmaceutical sectors. *Ind. Crops Prod.* **2016**, *83*, 329–338. [CrossRef]
13. Catalán, L.; Alvarez-Ortí, M.; Pardo-Giménez, A.; Gómez, R.; Rabadán, A.; Pardo, J.E. Pistachio oil: A review on its chemical composition, extraction systems, and uses. *Eur. J. Lipid Sci. Technol.* **2017**, *119*. [CrossRef]
14. Rabadán, A.; Álvarez-Ortí, M.; Gómez, R.; Alvarruiz, A.; Pardo, J.E. Optimization of pistachio oil extraction regarding processing parameters of screw and hydraulic presses. *LWT Food Sci. Technol.* **2017**, *83*, 79–85. [CrossRef]
15. Rabadán, A.; Álvarez-Ortí, M.; Pardo, J.E.; Gómez, R.; Pardo-Giménez, A.; Olmeda, M. A comprehensive approach to pistachio oil production. *Br. Food J.* **2017**, *119*, 921–933. [CrossRef]
16. Álvarez-Ortí, M.; Quintanilla, C.; Sena, E.; Alvarruiz, A.; Pardo, J.E. The effects of a pressure extraction system on the quality parameters of different virgin pistachio (*Pistacia vera* L. var. Larnaka) oils. *Grasas y Aceites* **2012**, *63*, 260–266. [CrossRef]
17. Sorho, S.; Yaya, S.; Augustin, A.A.; Laurent, L. Fatty acids and unsaponifiable composition of *Cucumis amaris* seeds oil. *J. Appl. Sci.* **2006**, *6*, 3167–3169.
18. Clifton, P.M.; Keogh, J.B. A systematic review of the effect of dietary saturated and polyunsaturated fat on heart disease. *Nutr. Metab. Cardiovasc. Dis.* **2017**, *27*, 1060–1080. [CrossRef]

19. Peh, H.Y.; Tan, W.S.D.; Liao, W.; Wong, W.S.F. Vitamin E therapy beyond cancer: Tocopherol versus tocotrienol. *Pharmacol. Ther.* **2016**, *162*, 152–169. [CrossRef]
20. Górnaś, P.; Soliven, A.; Segliņa, D. Seed oils recovered from industrial fruit by-products are a rich source of tocopherols and tocotrienols: Rapid separation of α/β/γ/δ homologues by RP-HPLC/FLD. *Eur. J. Lipid Sci. Technol.* **2015**, *117*, 773–777. [CrossRef]
21. Sen, C.K.; Khanna, S.; Roy, S. Tocotrienols in health and disease: The other half of the natural vitamin E family. *Mol. Asp. Med.* **2007**, *28*, 692–728. [CrossRef] [PubMed]
22. EEC. Commission Regulation (EEC) No 2568/91 of 11 July 1991 on the Characteristics of Olive oil and Olive-Residue Oil and on the Relevant Methods of Analysis. *Off. J. Eur. Communities* **1991**, *34*, L248.
23. Santos, J.; Álvarez-Ortí, M.; Sena-Moreno, E.; Rabadán, A.; Pardo, J.E.; Oliveira, M.B.P.P. Effect of roasting conditions on the composition and antioxidant properties of defatted walnut flour. *J. Sci. Food Agric.* **2018**, *98*, 1813–1820. [CrossRef] [PubMed]
24. Alves, R.C.; Casal, S.; Oliveira, M.B.P.P. Determination of vitamin e in coffee beans by HPLC using a micro-extraction method. *Food Sci. Technol. Int.* **2009**, *15*, 57–63. [CrossRef]
25. CIE. *Colorimetry*, 2nd ed.; Central Bureau of the Commission Internationale de L'Eclairage: Wien, Austria, 1986.
26. Alimentarius, C. Standard for Olive Oils and Olive Pomace Oils. In *Codex Alimentarius Commission*; CODEX STAN 33-1981; Food and Agriculture Organization of the United Nations (FAO): Rome, Italy, 1981.
27. Rabadán, A.; Pardo, J.E.; Gómez, R.; Álvarez-Ortí, M. Influence of temperature in the extraction of nut oils by means of screw pressing. *LWT* **2018**, *93*, 354–361. [CrossRef]
28. Rabadán, A.; Gallardo-Guerrero, L.; Gandul-Rojas, B.; Álvarez-Ortí, M.; Pardo, J.E. Effect of roasting conditions on pigment composition and some quality parameters of pistachio oil. *Food Chem.* **2018**, *264*, 49–57. [CrossRef]
29. Khachik, F.; Beecher, G.R.; Holden, J.; Tenorio, M.D.; Barrera, M.R.; Goli, M.B.; Lusby, W.R. Effect of Food Preparation on Qualitative and Quantitative Distribution of Major Carotenoid Constituents of Tomatoes and Several Green Vegetables. *J. Agric. Food Chem.* **1992**, *40*, 390–398. [CrossRef]
30. Kirk, J.; Tilnet-Basset, R. The plastids: Their chemistry, structure, growth and inheritance. In *Carotenoids and Phycobiliproteins*; Elsevier/North-Holland Biomedical Press: Amsterdam, The Netherlands, 1978.
31. Rabadán, A.; Álvarez-Ortí, M.; Gómez, R.; Pardo-Giménez, A.; Pardo, J.E. Suitability of Spanish almond cultivars for the industrial production of almond oil and defatted flour. *Sci. Hortic.* **2017**, *225*, 539–546. [CrossRef]
32. Pardo, J.E.; Cuesta, M.A.; Alvarruiz, A.; Granell, J.D.; Álvarez-Ortí, M. Evaluation of potential and real qualities of virgin olive oil from the designation of origin (DO) "Aceite Montes de Alcaraz" (Albacete, Spain). *Food Chem.* **2011**, *124*, 1684–1690. [CrossRef]
33. Schwartz, H.; Ollilainen, V.; Piironen, V.; Lampi, A.M. Tocopherol, tocotrienol and plant sterol contents of vegetable oils and industrial fats. *J. Food Compos. Anal.* **2008**, *21*, 152–161. [CrossRef]
34. Crews, C.; Hough, P.; Godward, J.; Brereton, P.; Lees, M.; Guiet, S.; Winkelmann, W. Study of the main constituents of some authentic walnut oils. *J. Agric. Food Chem.* **2005**, *53*, 4853–4860. [CrossRef] [PubMed]
35. Cuesta, A.; Álvarez-Ortí, M.; Pardo-Giménez, A.; Gómez, R.; Rabadán, A.; Pardo, J.E. Walnut virgin oil: A different but high quality vegetable oil. *Riv. Ital. Delle Sostanze Grasse* **2017**, *94*, 9–17.
36. Górnaś, P.; Siger, A.; Juhņeviča, K.; Lacis, G.; Šne, E.; Segliņa, D. Cold-pressed Japanese quince (Chaenomeles japonica (Thunb.) Lindl. ex Spach) seed oil as a rich source of α-tocopherol, carotenoids and phenolics: A comparison of the composition and antioxidant activity with nine other plant oils. *Eur. J. Lipid Sci. Technol.* **2014**, *116*, 563–570. [CrossRef]

37. Górnaś, P.; Siger, A.; Segliņa, D. Physicochemical characteristics of the cold-pressed Japanese quince seed oil: New promising unconventional bio-oil from by-products for the pharmaceutical and cosmetic industry. *Ind. Crops Prod.* **2013**, *48*, 178–182. [CrossRef]
38. Tasan, M.; Gecgel, U.; Demirci, M. Effects of storage and industrial oilseed extraction methods on the quality and stability characteristics of crude sunflower oil (*Helianthus annuus* L.). *Grasas y Aceites* **2011**, *62*, 389–398. [CrossRef]

foods

MDPI

Review

Review about Non-Lipid Components and Minor Fat-Soluble Bioactive Compounds of Almond Kernel

José M. Roncero [1,*], Manuel Álvarez-Ortí [1], Arturo Pardo-Giménez [2], Adrián Rabadán [1] and José E. Pardo [1]

1 Higher Technical School of Agricultural and Forestry Engineering, University of Castilla-La Mancha, Campus Universitario, s/n, 02071 Albacete, Spain; manuel.alvarez@uclm.es (M.Á.-O.); adrian.rabadan@uclm.es (A.R.); jose.pgonzalez@uclm.es (J.E.P.)

2 Mushroom Research, Experimentation and Service Centre, C/Peñicas, s/n, Apartado 63, Quintanar del Rey, 16220 Cuenca, Spain; apardo.cies@dipucuenca.es

* Correspondence: jmroncero@albacete.uned.es

Received: 7 October 2020; Accepted: 8 November 2020; Published: 11 November 2020

Abstract: This work presents a bibliographic review about almond kernel non-lipid components, in particular about the protein fraction, the carbohydrates and the mineral fraction. In addition, other fat-soluble phytochemicals which are present in minor concentrations but show important antioxidant activities are reviewed. Almond kernel is a rich protein food (8.4–35.1%), in which the globulin–albumin fraction dominates, followed by glutelins and prolamins. Within the almond kernel protein profile, amandine dominates. Free amino acids represent a small amount of the total nitrogen quantity, highlighting the presence of glutamic acid and aspartic acid, followed by arginine. Carbohydrates that appear in almond kernels (14–28%) are soluble sugars (mainly sucrose), starch and other polysaccharides such as cellulose and non-digestible hemicelluloses. Regarding the mineral elements, potassium is the most common, followed by phosphorus; both macronutrients represent more than 70% of the total mineral fraction, without taking into account nitrogen. Microminerals include sodium, iron, copper, manganese and zinc. Within the phytochemical compounds, tocopherols, squalene, phytosterols, stanols, sphingolipids, phospholipids, chlorophylls, carotenoids, phenols and volatile compounds can be found.

Keywords: tree nuts; chemical composition; proteins; carbohydrates; minerals; phytochemicals; polyphenols; antioxidants; volatile compounds

---

1. Introduction

The almond is the most cultivated nut in the world, where the estimated annual production exceeds 3 million tons [1]. Most of the world's production is concentrated in three regions, which include California, the Mediterranean Basin and the Middle East, although almond cultivation is also increasing in the Southern Hemisphere, in countries such as Australia or Chile.

Almond tree, *Prunus dulcis*, belongs, taxonomically, to the *Amygdalus* subgenus inside the *Prunus* genus, the *Rosaceae* family and the order *Rosales* [2]. Its cultivars are classified depending on the hardness of the shell. Soft and medium-hard shell cultivars, like Non Pareil and Guara, respectively, show low resistance to attacks by pests and are more susceptible to rancid oxidation, but show high kernel yields (55% and 35–40%, respectively) [3]. On the other hand, hard shell varieties present the lowest kernel yield (<25%), but they maintain in a better way the organoleptic and commercial characteristics, highlighting the importance of Marcona and Desmayo Largueta cultivars. Physical parameters are useful for cultivar determination even when the nuts are grown in the same conditions.

From the botanic point of view, the almond tree nut is a drupe. It is formed by the evolution of the ovary walls, which develop into the pericarp (hull), an outer layer that is formed by a pulpy and very fibrous tissue, that can be divided into the exocarp (thin and pubescent) and the mesocarp (thickest); and a lignified interior layer that creates a heavy to less heavy coat, the endocarp (shell). At maturity, the pulpy mesocarp dries and opens by its ventral suture, releasing the lignified endocarp. The seed, which constitutes the edible kernel and the commercial part of the nut, occupies the inner part, surrounded by the endocarp. The kernel contains the embryo coated by the teguments [2].

Almond consumption has been found to be associated with many health benefits [4], especially related to the reduction of the cardiovascular diseases risk, but also with effects on other pathologies, such as hypertension, diabetes mellitus or metabolic syndrome. These activities are generally attributed to the lipid fraction, where the fatty acid profile has a predominant role, but also minor compounds such as polyphenols and phytosterols may be involved. Moreover, recent studies have explored the effect of other nutritional compounds like fiber on gut microbiota [5] or the antioxidant capacity of the protein fraction [6].

Regarding the chemical composition of almond kernels, the fatty acid profile has been extensively studied and characterized. However, the information about other minor compounds and the non-lipid fraction, in which a large quantity of nutrients are found, is more scarce, and is generally presented separately, so it is difficult to find those gathered to obtain an overall view of the content of all these compounds in the almond kernel. Therefore, this review aims to present data on the composition of the non-lipid fraction as well as other less studied minor compounds in almond kernels, to provide an overview of all these compounds with potential benefits on human health.

2. Chemical Composition of Almond Kernel

The main fractions that can be found in almond kernels, other than water, are the lipid fraction, the protein fraction, carbohydrates and the mineral fraction. A numerous group of compounds called phytochemicals should also be added, because even though they appear in low quantities, they have a main role in almond quality. The proportion of these compounds changes according to the cultivars, the cultivation system and the geographical origin [7–12] (Figure 1).

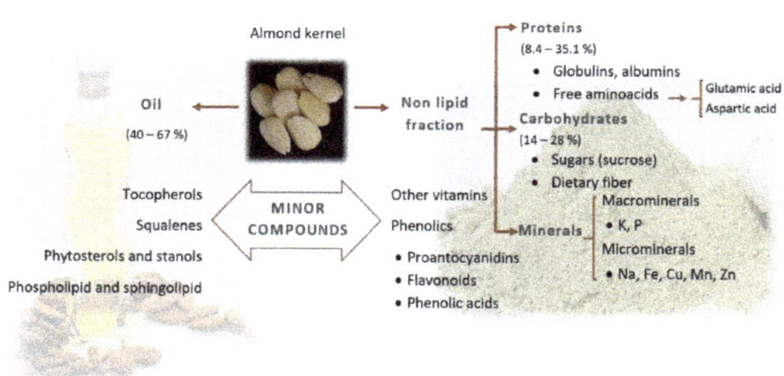

Figure 1. Chemical composition of almond kernel.

Precise knowledge about almond kernel composition is of great interest from the commercial, industrial and nutritional points of view, especially taking into account the variability that exists between different cultivars. In Yada et al. [8], total lipid values between 40 and 67 g/100 g of dry almond weight and between 35 and 66 g/100 g of almond fresh weight (f.w.) were reported. Almond oil is mainly composed of mono- and diunsaturated fatty acids [13]. In the case of total proteins (considering a conversion N factor of 5.18), the values oscillated between 14 and 61 g/100 g of almond fresh weight, and in the case of soluble sugars, values between 1.8 and 7.6 g/100 g of dry almond and between 2.5 and 12 g/100 g of fresh almond have been reported.

Regarding phenotypic correlations, a negative correlation was found between the oil and the total protein content [14]. This interdependence can be explained biochemically, because both fractions are formed during the ripening process from carbohydrates, which are abundant in the early stages of seed development but decrease over the ripening process [15].

In the existing literature, a clear evolution of the topic treatment can be observed. The first works about the chemical composition of almond kernels started appearing in the 1950s [16,17], providing data about the main fractions, without discrimination between cultivars and origins. In the decades of the 1970s and 1980s, works about the chemical composition appeared, referring to defined cultivars and providing information about fatty acids, amino acids, mineral salts and soluble sugars compositions [15,18,19]. Works studying the influence and effect of different labor systems, the place of origin and the harvest year on almond chemical composition also appeared.

From 1990, a step forward can be observed, related to the use of advanced statistical treatments, in such a way that not only composition data are given, but it is also tried to bring together genotypes that have a similar response and present close values to commercial effects [7,20–22]. In addition, food origin determination appears as a main target in food quality control and safety [23].

In the new century, the approach to the chemical composition of almond kernels, which can be applied to the rest of nuts, is focused on minority components, called phytochemicals. It has been shown that almond kernels present a wide range of substances with high nutritional value or with effects on health on one side [24–26] and with antioxidant effects on the other [27–29]. The interest aroused by these substances has boosted the development of new methods for their determination, increasing exponentially the published articles about them. Another important source of data is due to the recent interest in almond oil extraction as virgin edible oil [11–13,30–35]. In this sense, almond oil has been widely characterized, but oil extraction industries generate a by-product derived from the grinding of the pressing cake, which originates a partially defatted flour where the non-lipidic fraction takes on a special relevance. These flours have been reported to have promising uses in the culinary industry to enhance the nutritional properties of various products [36,37], or in mushroom cultivation, where it can be added as a nutritional supplement [34].

3. Protein Fraction of Almond Kernel

Almond kernel is a protein-rich food (second fraction in importance after the lipid fraction), but its content presents differences depending on the cultivar, weather conditions and cultivation area [9,14,16,32,34,38–42].

Table 1 shows the protein content of almond kernel samples with different origins found in relevant published articles. The percentage of variation ranges from 8.4%, found in Spanish samples [14], to 35.1%, found in Moroccan samples [42]. The differences in the protein content found in different samples may be related to the methods used in the analysis. To calculate the protein content, typically a specific conversion factor of nitrogen to protein of 5.18 is used [43], since amandine, which is the dominant protein in almond, is a globulin that contains 19.3% of nitrogen [44]. However, other studies use the general conversion factor (6.25), based on the nitrogen content of most common proteins, which could lead to overestimate the protein content. This point could explain some discrepancies found within the data. For this reason, data regarding total nitrogen would be more useful to compare samples from different origins.

Table 1. Macronutrients content (%) of almond samples with different origins.

Nutrient	Range of Variability (g/100 g)	Origin	Source of Variability Studied			References
			V *	E **	Ap ***	
Protein, total (N × 5.18)						
	16.4–22.1	USA	-	-	-	[38]
	18.5–24.0	California	Yes	Yes	-	[9]
	20.7–23.3	USA	Yes	-	-	[45]
	15.8–25.1	Spain	Yes	Yes	-	[18]
	14.5–29.2	Spain	Yes	Yes	-	[39]
	8.4–24.7	Spain	Yes	-	-	[14]
	14.1–26.5	Spain	-	-	-	[3]
	21.0–24.0	Portugal	Yes	Yes	-	[46]
	9.6–28.5	France, Italy and Greece	Yes	Yes	-	[41]
	20.0–32.8	Spain and Morocco	Yes	Yes	-	[39]
	14.1–35.1	Morocco	Yes	Yes	-	[42]
	16.7–31.5	Turkey	Yes	Yes	-	[47]
	12.7–16.3	Turkey	Yes	-	-	[40]
	20.4–25.8	Turkey	Yes	Yes	-	[48]
	11.52 ± 1.1	Nigeria	-	-	-	[49]
	23.8	India	-	-	-	[50]
	20.0	South Africa	-	-	-	[51]
	17.36–23.02	Serbia	Yes	-	Yes	[52]
Carbohydrates, total						
	14–21	Portugal	Yes	Yes	-	[46]
	23.6–27	USA	Yes	-	-	[45]
	28	Nigeria	-	-	-	[53]
	28.0	South Africa	-	-	-	[51]
Sugars, soluble						
	2.6	Turkey	Yes	-	-	[54]
	7.9	Spain	Yes	-	-	[15]
	1.74–4.31	Greece	Yes	-	Yes	[55]
Sucrose						
	2.5–5.1	California	Yes	Yes	-	[9]
	1.42–3.62	Greece	Yes	-	Yes	[55]
	1.15–2.22	Portugal	Yes	-	-	[56]
	3.67–7.09	Spain	-	-	Yes	[57]
	1.21–3.08	Portugal	Yes	-	-	[58]
Starch						
	0.4–1.4	Italy	-	-	-	[59]
Fiber, total dietary						
	9.8	California	-	-	-	[60]
	7.9–16	California	Yes	Yes	-	[9]
	3.3–8.6	Spain	Yes	Yes	-	[22]
	4.73–6.01	Spain	-	-	Yes	[57]
	11–14	Italy	-	-	-	[59]

* Variety; ** Environment/crop year; *** Agronomic practices (irrigation, fertilization, etc.).

Font i Forcada et al. [61] found that two quantitative trait loci (QTL) controlled the total protein content. The first marker LG6, located in the lowest part of the almond linkage groups, had a logarithm of the odds (LOD) values of 3.21 and explained a phenotypic variance of 17%. The second QTL was found in the lowest part of LG7 and had a similar effect, with an LOD of 3.18 explaining a phenotypic variance of 16.6%.

Nitrogen total content of almond samples has shown different percentages: 3% [15], 4.06% [62], 4.23% [54] and 4.62% [63].

3.1. Protein Profile

Proteins are classified depending on their solubility in albumins (soluble in water and dilute solutions), globulins (classified into euglobulins—soluble in dilute solutions, acids and alkalis and insoluble in water, and pseudoglobulins—moderately soluble in these solutions), prolamins (soluble in solutions with 50–90% ethanol), glutelins (soluble in dilute acids and alkalis) and scleroproteins (insoluble in all mentioned solvents).

Saura et al. [15] found that the protein dominant fraction was that composed by globulins–albumins, with about 90% in all studied samples. On the other hand, glutelins accounted for between 4 and 11% of total proteins, while prolamins were found below 0.4% in all cases.

Within the protein profile, amandine is dominant, also known as the almond major protein (AMP), which represents 65% of total proteins of almond that can be extracted in aqueous medium [64]. This protein is the main component responsible for food allergies caused by almonds, due to the antigenic activity that it presents. It is an ideal marker to detect traces of almond in foods [24].

3.2. Free Amino Acids

Together, free amino acids (AAs) represent a small quantity of total nitrogen matter (1%), which matches with the low content of non-protein nitrogen; consequently, the total amino acids content in almond kernel is a good approximation of the total protein content. The main free amino acids found in almond kernel are glutamic acid and aspartic acid (including glutamine and asparagine), followed by arginine. Phenylalanine, alanine, serine and threonine are also present although with lower quantities [15,65]. Amrein et al. [66] also found that aspargine was the main free amino acid in raw almond kernel (20–50% of total free amino acids).

The free amino acids fraction has been used for the characterization of almond cultivars [65,67]. These free amino acids are important due to their contribution to food taste and for being precursors of aromatic components and colored substances that are produced during the obtention, preparation and storage of food.

Esteban [18] found a higher content in almost all AAs in cultivars grown in northwest Spain compared to those grown in the southwest, which was related to the lower content of fats in the northwest cultivars.

3.3. Essential Amino Acids

Humans are unable to synthesize eight AAs that need to be necessarily obtained through diet, and these are known as essential amino acids. Arginine and histidine must be included in this group with essential amino acids (threonine, methionine, valine, isoleucine, leucine, phenylalanine, tryptophan and lysine) because they are considered essential for children but not for adults. In addition, cysteine and tyrosine are considered semi-essential amino acids, due to the sparing effect they have on methionine and phenylalanine, respectively.

Esteban [18] and, later, Saura et al. [15] found that the most abundant essential amino acid in almond kernel was arginine, with an average value of 524 mg/g of N. Essential AAs represent 28% of total amino acids. Regarding the limiting amino acids in almond kernel, comparing to egg, they found that the first one would be lysine, followed by threonine. Ahrens et al. [45] found that methionine together with cysteine was the limiting amino acid, followed by lysine and threonine. More recently,

House et al. [68] considered that Ahrens et al. had underestimated the total sulfur AA content due to a method issue. They found lysine as the limiting AA.

On the other side, the inhibitory activity of trypsin was evaluated, as well as the hemagglutinating activity, not being detected in analyzed samples.

The digestibility of the protein and ultimate utilization of the constituent AAs for metabolic functions are equally important in assessing protein quality [68]. Amino acid score (AAS) together with digestibility is a parameter that allows calculating the protein digestibility-corrected amino acid score (PDCAAS), which can be used to properly establish the protein quality index. In this sense, although almond kernel proteins show a high degree of digestibility, higher than 80% in all analyzed samples measured as protein digestibility in vivo, when the protein digestibility index is corrected by the AAS, it results in low quality [45]. On the other hand, House et al. [68] obtained better values. This kind of study has not been contemplated in the rest of the previously cited research.

4. Carbohydrates in Almond Kernel

The carbohydrates from almond kernel are soluble sugars, starch and other polysaccharides such as celluloses and hemicelluloses that are non-digestible, but they have physical effects in the intestinal tract with benefits for human health [8]. The total carbohydrates content ranged from 14% to 28% (Table 1). The sugars that can be found in almond kernel, although not found in high concentrations, are enough to provide the sweet flavor to almonds.

4.1. Sugars

Nuts are characterized by low quantities of soluble sugars, that range from 2.6% to 7.9% (Table 1). The soluble sugars fraction suffers important quantitative variations depending mainly on the cultivar, but also on the origin and even harvest time [9,18,69,70]. Most sugars are not reducing, with sucrose representing more than 90% of the total sugars. Raffinose, inositol, sorbitol, fructose and glucose were also detected [55,56,71]. Differences found between commercial and regional cultivars about sugars composition, especially sucrose, may be used to establish a sugar profile as an indicator of almond kernel quality.

Moreover, Nanos et al. [69] found galactose (reducing sugars), melezitose (tri-saccharide) and stachyose (tetrasaccharide). Late-harvested almonds had lower total amounts of these sugars than early-harvested ones.

The fact that the main sugar is sucrose is due to its preferential production and its accumulation in almond kernel during ripening and to the fact that many minor sugars constitute a substrate for the synthesis of sucrose [57]. Kazantzis et al. [55] observed that almond kernels early-harvested (green mesocarp surface at 90%) had a lower content in sucrose and higher content in inositol than more ripened almonds (brown mesocarp surface at 90%). Trees subjected to irrigation and compost treatments also produce almond kernels with the highest content in sucrose and glucose. This could reduce their water content, causing a higher concentration of compounds such as sugars [57]. In contrast, a recent study by Lipan et al. [72] showed that almost all morphological and physical-chemical parameters were unaffected by water stress.

Regarding the total sugars, a value of 4.63 g/100 g of peeled almonds was found, distributed as 4.46 g of sucrose, 0.03 g of glucose and 0.14 g of maltose [43]. A review of a worldwide collection of almond samples found a soluble sugars content ranging from 1.80 to 7.60 g/100 g f.w. [8]. On the other side, when data are referred to almonds with skin, total values fall to 4.35 g/100 g, and are distributed as follows: 3.95 g of sucrose, 0.17 g of glucose and 0.04 g of maltose, appearing with 0.11 g of fructose and 0.07 g of galactose.

4.2. Starch

Although starch is the main reserve carbohydrate in many fruits and seeds, in almond kernel, it does not reach remarkable values. Thus, this compound has not attracted much attention among researchers.

Ruggeri et al. [59] found a percentage of 0.4% to 1.4%. The Agriculture Department of the U.S.A. indicated a starch value of only 1.0 g/100 g of skin peeled almond kernels [43].

4.3. Fiber

Fiber is a heterogeneous mix of polysaccharides (cellulose, hemicelluloses, gums and mucilages and pectin substances) and non-polysaccharides (lignin, non-digestible proteins and other). Terms commonly used to define it are as follows: crude fiber, composed of cellulose (50–80%), hemicellulose (~20%) and lignin (10–50%); neutral detergent fiber (NDF), consisting of cellulose, hemicellulose and lignin; acid detergent fiber (ADF), consisting of cellulose and lignin; and acid detergent lignin (ADL) [73]. On the other side, dietary fiber can be defined as a group of components that are not digested by enzymes in the human gastro-intestinal tract, and as being mainly composed of cellulose, hemicellulose, lignin, pectin and non-digestible proteins.

The content in neutral detergent fiber, whose fundamental components are cellulose, lignin and hemicellulose, provides the value of dietary fiber, while acid detergent fiber, whose fundamental components are cellulose and lignin, provides the crude fiber value.

Barreira et al. [46] reported data of neutral detergent fiber between 2.9% and 3.2%, depending on the selected cultivar. However, dietary fiber content in almonds ranged from 3.3% to 16% (Table 1). Kodad [22] calculated a variation coefficient of 9.81% for dietary fiber. Significant differences between selected cultivars, years and the interaction "genotype" × "year" have been found regarding dietary fiber content, which confirms the large variability of this character between genotypes and years. Yada et al. [9] found that the effect of harvest year on dietary fiber was highly significant ($P < 0.01$), to the point of having blocked the observation of cultivar differences. In addition, some other agronomic treatments may have an effect on the content of dietary fiber. For example, when organic fertilizers were applied, higher fiber content was observed in the fruit than when an inorganic fertilizer was employed [57].

Crude fiber concentration in almond kernel also shows high variability. First references indicated low crude fiber contents, about 2% or 3% of dry matter [15,16], while other recent results reached contents of 5.81% [22], probably due to improvements in analytical techniques.

5. Mineral Fraction of Almond Kernel

5.1. Ashes

Mineral content is sometimes expressed as the ash content, which is the inorganic residue that remains after the incineration of the plant tissues. Almond kernels contain approximately 3 g ash/100 g of fresh weight [74,75]. These values may vary depending on the study considered (Table 2), between 2.3% [9] and 5.0% [51].

Table 2. Average value or range of main mineral elements (macro- and microminerals) found in almond kernel in the literature (mg/100 g).

Ash (g/100 g)	K	P	Ca	Mg	S	Cl	Na	Fe	Cu	Mn	Zn	Origin	Reference
	435	577	298	299	587		2.27	3.4	0.96	1.36	3.04	Spain	[7]
2.69–3.6	821	585	275	281	130	14	10.8	4	1.2	1.6	3.8	Spain	[15]
	618–785	345–507	88–124	242–285			4.7–15.5	3.5–5.3	1–1.6	1.1–1.7	3.4–3.9	Spain	[18]
2.74–3.05	1373.8	873.8	243.2	351			32.6	23.4	1		5	Turkey	[40]
	1546–1685	253–259	640–678	447–494				5.5–6.5	2.4–2.6	3.8	7.6–8.0		[63]
	1050	300	467	30				7.0	0.5		3.4	Italy	[74]
3.03–4.66	1677–2051	404–800	98–187	361–513			5.66–10.38	3.98–14.6	1.60–2.30	2.90–3.39	7.78–8.84	Turkey	[54]
-	465–1235	119–748	160 -663	100–333								France, Italy and Greece	[41]
2.3–3.4	543–902	364–548	198–373	224–303				2.58–4.47	0.46–1.57	1.31–3.98	2.02–4.03	California	[9]
3.29–4.66	679–986	584–697	250–332	325–381			9.20–16.06	6.08–10.62	2.02–3.97	2.52–4.76	4.80–9.53	Turkey	[48] *
5.0			539.2	542.4				7.15	2.37	2.58	4.97	South Africa	[51]
		133.25	450.0					6.25				India	[76]

* Referred to non-dried matter. K: potassium; P: phosphorus; Ca: calcium; Mg: magnesium; S: sulfur; Cl: chloride; Na: sodium; Fe: iron; Cu: copper; Mn: manganese.

The sum of mineral elements is sensibly lower than the ash content, with percentages around 60%, which is fundamentally explained because the oxygen associated with these minerals is not counted in the ashes obtained by calcination [15]. According to Esteban [18], the percentge of all minerals, excluding nitrogen, represents between 51.3% and 55.2% of total ash content.

5.2. Macrominerals

Macrominerals refer to those minerals that are needed in quantities higher than 100 mg/day. On the other hand, those that are needed in small quantities are called microminerals, oligo elements or trace elements. Table 2 shows the average value of the main mineral elements (macrominerals and microminerals) found in almond kernel.

Potassium is the major element in all studies, except the one carried out by Prats [7], followed by phosphorus. Both elements represent 70% of the mineral fraction, not counting nitrogen. The next in importance are calcium and magnesium with very close values, in such a way that in some samples, one is higher and in others the opposite happens [15,18]. Globally, the mean magnesium values are higher than calcium values, and both represent half the phosphorus content, or even less [15].

Sulfur also appears in high amounts, although it is an element that is not commonly analyzed in comparison with the previous ones. Its values vary greatly depending on the study, probably due to the different methods applied for its determination. Prats [7] found higher values, comparable to phosphorus values. Macronutrients aggregation, not counting nitrogen, represents large percentages which are almost identical between cultivars, ranging from 98.0% to 98.7% of total minerals. Among Chinese wild almond species, potassium contents between 534 and 663 mg/100 g, calcium contents between 80 and 229 mg/100 g and magnesium contents between 194 and 239 mg/100 g have been found [77].

5.3. Microminerals or Trace Elements

Main microminerals or trace elements found in almond kernel are sodium, chlorine, iron, copper, manganese and zinc (Table 2). Sodium and chlorine are those that appear in higher proportion [15,16,54], followed by iron and zinc contents, which also show important values. In this case, as it happened with calcium and magnesium, for some authors, the content of iron is higher, and for others, the zinc content, but generally the quantity, is lower than 5.5 mg/100 g. Nevertheless, attention should be paid to the high contents in iron and zinc found by Ozcan et al. [40] and Aslantas et al. [54], respectively. Among Chinese wild almond species, iron contents between 4.6 and 6.0 mg/100 g and zinc contents between 4.1 and 5.6 mg/100 g have been found [77].

Other elements found in almond kernel, although in minor concentrations, include molybdenum that ranges from 4 to 30 μg/100 g, boron which ranges between 0.18 and 2.9 mg/100 g [15,16,78], chromium ranging between 0.04 [79] and 0.17 mg/100 g [78], aluminum ranging between 0.83 [79] and 2.2 mg/100 g [78], nickel with 0.034 mg/100 g [79] and selenium with 0.004 mg/100 g [51].

Some references to toxic heavy metals have also been found [50,51,79]. Even though some heavy metals such as cobalt, copper, chromium, manganese and nickel are needed for humans in small proportions, others may be carcinogenic or toxic, affecting the central nervous system (manganese, mercury, lead, arsenic), kidney or liver (mercury, lead, cadmium, copper), or the skin, bones or teeth (nickel, cadmium, copper, chromium).

6. Phytochemical Compounds of Almond Kernel

Phytochemicals, also known as bioactive compounds, are mainly additional nutritional compounds that can be found in certain foods, and that show an important and interesting physiological activity with positive effects on human health, which makes them very valuable elements for the scientific community and the food industry.

Several thousands of phytochemicals have been reported, some of them having a strong antioxidant activity (catechin, quercetin, tannin, ellagic acid, chlorogenic acid, cyanidin, etc.) [80], which are added to the already known antioxidant nutrients (vitamins A, C, E, selenium, etc.).

Phytochemicals comprise the following chemical groups: carotenoids, phenolic compounds, organosulfur compounds, some nitrogen compounds and alkaloids. Bolling et al. [81] added a carbohydrates group to this classification, the phytates, and together with the carotenoids, they include other unsaponifiable compounds of the lipid fraction.

6.1. Tocopherols (Vitamin E)

Tocopherols, or vitamin E, are a group of soluble compounds that includes four tocopherols (designated as α, β, γ and δ) and four tocotrienols (designated as α, β, γ, y and δ) [80]. Tocopherols are natural mono-phenolic components with different antioxidant activity, which have several homologues depending on the position and number of methyl radicals. Their main biochemical function is probably the protection of polyunsaturated fatty acids against peroxidation. A good number of scientific studies focused, in the first instance, on the tocopherol content and its effect on the maintenance of oil properties. Almond kernel is considered one of the richest foods in α-tocopherol [82,83].

Tocopherol content in almonds shows a wide range of variability, as summarized in Table 3. The form with higher concentration in almond kernel oil is α-tocopherol. Variability depends on almond genotypes (cultivars), climatic conditions and environmental conditions. Kodad et al. [10], in a study about 44 Spanish cultivars, for two consecutive years, found a large variability in tocopherol concentrations, in almond oil, with a significant effect of both the genotype, the year and the interaction genotype × year. The main source of variability appeared due to the genotype. The geographical origin was significant with higher concentrations of tocopherols in almond populations with a mountainous origin, probably due to the empiric selection to increase the shelf life, since tocopherol retards the rancidity appearance. Abiotic stress leads to higher tocopherol contents due to its protective role. Similar conclusions reached Zhu [28], after analyzing samples of cultivars from Australia, Spain and the United States, and Yada et al. [9], with Californian cultivars from different regions. Besides, as the trees matured from one year to the next, the vitamin E concentration increased [10]. The obtained results also show that the homologues α and δ are those that present higher variability. Higher concentrations found by Maestri et al. [30] in the Argentine Northeast, where the kernel development matches mainly with spring and summer, with medium temperatures that are warmer to those typically observed in the Mediterranean region, can explain these values.

Table 3. Vitamin content in almonds.

Nutrient	Range of Means		Origin	Variability Sources			References
	mg/100 g Almonds	mg/100 g Almond Oil		Variety	Environment	Extraction Method	
			Vitamin E homolog				
Total tocopherols		50.1–49.0	Spain	Yes	Yes	-	[12]
α-tocopherol		24.2	-	-	-	-	[84]
		17.4	-	-	-	-	[85]
		37.0–57.1	Argentina	Yes	Yes	-	[30]
		18.0–32.0	California	Yes	Yes	-	[9]
		30.9–65.7	Morocco	Yes	Yes	-	[42]
		31.3–54.6	Morocco	Yes	Yes	-	[86]

Table 3. *Cont.*

Nutrient	Range of Means		Origin	Variability Sources			References
	mg/100 g Almonds	mg/100 g Almond Oil		Variety	Environment	Extraction Method	
	34.9		California	-	-	-	[83]
	5.96–19.42		Portugal	Yes	-	-	[58]
		27–38	Portugal	Yes	Yes	-	[46]
		42.0–54.2	Spain	Yes	-	-	[35]
	23.7–37.4		Spain	Yes	-	-	[35]
		14.18–17.96		Yes	-	Yes	[87]
β -tocopherol		3.1	-	-	-	-	[84]
		1.7	-	-	-	-	[85]
		0.18–0.24	Portugal	Yes	Yes	-	[46]
ϒ-tocopherol		3.1	-	-	-	-	[84]
		5.7	-	-	-	-	[85]
		0.54–4.25	Morocco	Yes	Yes	-	[86]
		0.7–2.1	Portugal	Yes	Yes	-	[46]
	1.4		California	-	-	-	[83]
		0.67–2.79	Spain	Yes	-	-	[35]
	0.17–1.4		Spain	Yes	-	-	[35]
δ-tocopherol		n.d.	-	-	-	-	[84]
		1.7	-	-	-	-	[85]
		0.017–0.24	Morocco	Yes	Yes	-	[86]
		0.02–0.05	Portugal	Yes	Yes	-	[46]
α-tocotrienol		Traces	-	-	-	-	[85]
		0.04–0.2	Portugal	Yes	Yes	-	[46]
		0.3–0.5	Spain	Yes	-	-	[35]
ϒ -tocotrienol		0.11–0.24	Portugal	Yes	Yes	-	[46]
			Other vitamins				
Biotin	0.01–0.05		California	-	-	-	[17]
	0.12–0.90		Italy	-	-	-	[88]
Folate	0.10–0.13		California	-	-	-	[17]
Niacin (B3)	3.3–3.7		California	-	-	-	[17]
	1.5–3.4		Italy	-	-	-	[88]
		1.40–5.02	California	Yes	Yes	-	[9]
Pantotenic acid	0.36–0.38		California	-	-	-	[17]
Pyridoxine (B6)	0.16		California	-	-	-	[17]
	0.186		California	-	-	-	[83]
Riboflavin (B2)	1–1.1		California	-	-	-	[17]
		0.58–2.27	California	Yes	Yes	-	[9]
	1.432		California	-	-	-	[83]
Thiamine (B1)	0.19–0.25		California	-	-	-	[17]
	0.192		California	-	-	-	[83]

6.2. Vitamins

Most studies about vitamin content in almond have been focused on vitamins with an antioxidant effect, particularly vitamin E. However, almond kernels are a good source of vitamins B1 (thiamine), B2 (riboflavin), B6 (pyridoxine) and niacin (Table 3). Some kernel processing operations, like roasting or blanching (to a lesser extent), may result in vitamin loss due to the temperature effect on vitamin degradation [89].

6.3. Squalenes

Squalenes are polyunsaturated acyclic hydrocarbons with a triterpenoid lipophilic structure, similar to the vitamin E structure, and they contribute to the oxidation stability of vegetable oils because they prevent peroxidation of fats acting mainly against peroxyl radicals.

Squalene acts as a biosynthetic precursor to all steroids in plants and animals. However, Cherif et al. [90] detected a dramatic decrease in sterols at the 10–12th maturation week that suggested there was the absence of the synthesis of novo sterols from squalene which was maintained an enzymatic activity until the end of maturity. Squalene has important beneficial effects on health, such as decreasing the risk for various cancers and reducing serum cholesterol levels [91]. Squalene contents in almond oils ranged from 37.9 to 114.2 μg/g of oil (Table 4).

Table 4. Minor compounds: phytosterols, terpenic alcohols, squalene, aliphatic alcohols and tocopherols.

Compound	μg/g (%)	References
	Desmethylsterols	
Cholesterol	n.d.–7.18 (0.25)	[85,92–94]
24-Methylene-cholesterol	1.15 (0.04)–3.9	[93,94]
Campesterol	49–134 (2.46–16.7)	[11,12,85,92–95]
Campestanol	3.73 (0.13)-33	[94,96]
Stigmasterol	3.9–50 (0.41–6,9)	[85,92–96]
Δ7-Campesterol	22.39 (0.78)	[94]
Δ5,23-Stigmastadienol + Clerosterol	30.8–40.19 (1.40)	[93,94]
β-Sitosterol	580–2290 (72.4–95.5)	[11,12,85,87,92–96]
Sitostanol	32–54.83 (1.91)	[93,94,96]
Δ5-Avenasterol	32–283.89 (3.52–9.89)	[85,92–96]
Δ5,24-Stigmastadienol	6.05–42.48 (0.29–1.48)	[92–94]
Δ7-Stigamasterol	9.74–55.69 (0.43–1.94)	[12,93,94]
Δ7-Avenasterol	5.42–39.90 (0.26–1.39)	[11,93,94]
Total μg/g	1222–2870	[11,12,85,92–94,96,97]
	Methylsterols	
Obtusifoliol	7.90 (26.57)	[94]
Gramisterol	4.00 (13.46)	[94]
Citrostadienol	17.83 (59.96)	[94]
	Dimethylsterols	
Dammaradienol	0.62 (7.44)	[94]
Taraxerol	0.38 (4.55)	[94]
α + β Amyrin	2.08 (24.97)	[94]

Table 4. *Cont.*

Compound	μg/g (%)	References
Cycloartenol	0.99 (11.91)	[94]
24-Methylencycloartanol	4.27 (51.13)	[94]
Total sterols	2908.56	[94]
Squalene μg/g	37.9–114.2	[30,90,94,98]
	Terpenic alcohols	
Phytol	(71.65)	[94]
Geranylgeraniol	(28.35)	[94]
Total μg/g	9.74	[94]
	Aliphatic alcohols	
C22-OH	(23.13)	[94]
C23-OH	(2.56)	[94]
C24-OH	(29.66)	[94]
C25-OH	(7.70)	[94]
C26-OH	(40.31)	[94]
Total μg/g	5.55	[94]

In parenthesis: samples origin country/region; in brackets: source of variability studied. [a] variety; [b] environment/crop year; [c] extraction method. [94]: (Brazil); [92]: (Iran); [85]: (Sweden); [30]: (Argentina) [a,b]; [93]: (Turkey) [c]; [95]: [c]; [87]: (Turkey) [c]; [96]: (USA); [11]: (Spain) [a,b]; [12]: (Spain) [c]; [90]: (Tunisia) [a,b].

6.4. *Phytosterols and Stanols*

Phytosterols or plant sterols have a structure similar to cholesterol, while stanols are saturated sterols. They can be found in almond kernel, in free form or esterified with fatty acids [80]. β-sitosterol lowers cholesterol levels, enhances immunity and has anti-inflammatory, antipyretic and anti-carcinogenic effects (prostate essentially) [92].

Sterols are the most abundant class of compounds in the unsaponifiable matter. Desmethylsterols are the most commonly analyzed group, being β-sitosterol the main desmethylsterol with values of 95.5 % of total phytosterols [12], although with significant differences among genotypes [70,90,94]. As regards methylsterols, citrostadienol is the main compound, and regarding dimethylsterols, the total amount was around 30 μg/g [94].

β-sitosterol and campesterol are the dominant sterols in almond kernel. Δ5-Avenasterol, Δ7-Stigamasterol and stigmasterol are well represented in almond oils. Campestanol is the main stanol (Table 4). β-sitosterol is fundamentally found in almond kernel skin, while stigmasterol predominates in mesocarp [99].

Some studies have focused on the physiological phenomenon of phytosterols accumulation: biosynthesis and evolution, finding that the phytosterols amount depended on the harvest time [90,93]. Cherif et al. [90] found a relationship between the biosynthetic compounds of the glyceridic fraction of almond oil (mainly fatty acids) and those of the unsaponifiable fraction (particularly sterols). This relation may be established by 24-methylene cholesterol.

Ozcan et al. [87] compared varieties and extraction methods (cold press and Soxhlet methods). Both affected β-Sitosterol composition of the oil obtained. Neither extraction temperature nor extraction speed affected the total content of sterols in oils from the screw press, but higher temperatures caused a reduction in the content of Δ7-stigmasterol [12].

6.5. Sphingolipids and Phospholipids

Both components are polar lipids. Sphingolipids are complex lipids that are derived from sphingosine (unsaturated amino alcohol with 18 carbons), which is joined to a long-chain fatty acid by an amide bond forming the ceramide. Sphingolipids of plants are mainly cerebrosides (mono- and oligohexosilceramides) with a sugar molecule such as glucose, galactose, mannose and inositol. They are commonly found in cell walls, lipoproteins and other lipid-rich structures [99]. Phospholipids are a kind of lipid made up of a glycerol molecule, two fatty acids (1,2-diacylglycerol) and a phosphate group.

There are few studies about polar lipids in almond oil. Phospholipids and sphingolipids are the main classes of polar lipids with approximately 78% and 22%, respectively [100]. Between phospholipids, lecithin or phosphatidylcholine (45%), phosphatidylethanolamine or cephalin (45%), phosphatidylinositol (8%) and fosfatidiglycerol acid (2%) are the main compounds [101]. Fang [102] studied the sphingolipids content in almond kernels and found that the concentration of cerebroside (d18:2-C16:0h-glucose) was 0.068 mg/g of almond.

Compared with other nuts, almonds might not be the first choice for phospholipids, with relatively low compounds abundance and content [103]. Only 1.67% of the total fat is phospholipids, in comparison with 3.81% found in pistachios. The fatty acids of 16:0, 18:0, 18:1 and 18:2 are the most common structures of the fatty acyl moiety in almonds; phosphocholine, phosphoetanolamine and phosphoinositol are three major phospholipids species detected in almonds, representing 84% of total phospholipids.

6.6. Chlorophylls and Carotenoids

Tree nuts contain very low amounts of carotenoids [27]. Marginal pigments content has been found in wild almond kernels [92], and the only study that indicates the chlorophyll and carotenoid content for cultivated almond kernels is due to Ojeda-Amador et al. [35], who found 8.5–18 mg/kg of chlorophyll and 5.3–8.8 mg/kg of carotenoids in almond oils. Carotenoids concentration in almond kernels is low; consequently, it does not constitute an important dietary source of these substances [97].

6.7. Phenols

Phenols are the main phytochemical group and comprise the broad term "polyphenols", which are molecules with one or more phenolic groups and one or more hydroxyl groups and comprise a large and heterogeneous group of secondary plant metabolites. They are synthesized from carbohydrates and are generally produced as defense mechanisms against pathogens and the excess of ultraviolet radiation and to attract pollinators [104]. A general description of the biosynthetic pathways and regulation of phenolic compounds in stone fruits appears in the review by Lara et al. [105]. They are responsible for the sensorial and nutritional quality and antimicrobial, antiviral and anti-inflammatory properties are also attributed to them. Beyond antioxidant properties, they also have a variety of biological activities, including antioxidant, anti-inflammatory, vasodilatory and anticarcinogenic actions and also reduce cholesterol [82,97]. In recent years, there has been an increasing interest in biological properties of natural phenolic compounds as actors in the prevention of diseases in which oxidative stress reactions are involved [106].

In an extensive review, Bolling [107] reported a total phenolic content in whole almond that ranged from 0.47 to 13.40 mg/g gallic acid equivalents (GAE); meanwhile, skinless kernels varied between 0.64 and 0.71 mg/g (GAE). Approximately 130 different polyphenols have been identified in almond, although not all of these have been quantitated. Table 5 shows the range of variability found in the scientific literature about total phenolic content, total flavonoids content and total proanthocyanidins content, and Table 6 reflects the main quantitative results about phenolic compounds in almonds. Unpeeled almond kernels have a content of total phenols higher than peeled almond kernels [97,108]. The skin represents approximately 4% of the total almond weight and contains 70–100% of total phenols present in the nut [109]. The residual cakes could be expected to possess an added value for applications in food formulations since they are a good source of phenolic compounds that concentrate in the by-product due to their polar properties [35].

The phenolic content of almond skins depends on the industrial processing used. High temperatures (i.e., blanching, drying, roasting) could promote degradation of polymeric compounds such as proanthocyanidins, hydrolysis of glycosylated flavonoids and the decomposition of aglycones, which could explain the increase observed in the content of monomeric and oligomeric flavan-3-ols after drying or roasting, and the decline in flavonol and flavanone aglycones found after these treatments [110]. The total contents of phenolic compounds identified were significantly ($P < 0.05$) higher (about 2-fold) in the roasted samples than in the blanched almonds (freeze-dried). Roasting is the most suitable type of industrial processing of almonds to obtain almond skin extracts with the greatest antioxidant capacity [111].

Cultivars, climate and geography can affect total phenols concentration in almond kernels [57,58,112,113]. For Rabadán et al. [12,70], the variability of total polyphenol content depended mainly on the crop year. The use of pesticides reduces the phenols content, so it is advisable to implement organic production [114]. Among cold-pressed oils, the press system (screw or hydraulic) and the different extraction conditions considered did not generate significative differences [11,31].

Inside the phenol groups, mainly tannins, flavonoids, phenolic acids and stilbenes can be found.

Table 5. Total phenolic content (mg/g) gallic acid equivalents (GAE), total proanthocyanidins (mg/100 g) and total flavonoids (mg/100 g) in almonds, almond oil and defatted almond cake.

Range of Means				**Origin**	**Variability Sources**			**References**
Almonds	**Almond Skins**	**Almond Oil**	**Defatted Almond Cake**		**Variety**	**Environment**	**Extraction Method**	
Total phenolic content (mg/g) gallic acid equivalents (GAE)								
4.18				USA	-	-	-	[115]
1.27–2.41	0.099–0.268			California	Yes	-	-	[109]
1.30–4.56				Austria	Yes	-	-	[84]
0.45–0.49 [a]				Austria	Yes	-	-	[84]
1.10–2.90								[97]
0.09–1.63				Portugal	Yes	Yes	-	[112]
	27.1–59.1			Morocco	Yes	-	-	[113]
		0.019–0.022 [b]		Spain			Yes	[31]
		0.0085–0.0324		Spain	Yes	Yes	Yes	[11]
		0.019–0.026		Spain	Yes		Yes	[12]
0.03–0.81				Portugal	Yes	-	-	[58]
0.71–1.26		0.003–0.006	0.82–2.06	Spain	Yes	-	Yes	[35]
0.20–1.39				Serbia	Yes	-	-	[116]
Total proanthocyanidins (mg/g)								
	0.15–48.80			Spain	-	-	Yes	[117]
	5.81–28.80			Spain	-	-	Yes	[118]
0.70–2.90 [c]				-	-	-	-	[62]
	5.00–25.00			-	-	-	Yes	[119]
Total flavonoids (mg/g)								
6.24–25.02				Portugal	Yes	-	-	[112]

Table 5. Cont.

Range of Means				Origin	Variability Sources			References
Almonds	Almond Skins	Almond Oil	Defatted Almond Cake		Variety	Environment	Extraction Method	
	84.68–237.20 [d]			Portugal	Yes	-	-	[56]
	14.1–25.7			Morocco	Yes	-	-	[113]
n.d.–5.45								[120]
12.88–19.49				Portugal	Yes	-	-	[58]

a: blanched kernels without skin; b: caffeic acid equivalents; c: total tannins; d: hull extract; n.d.: not detected.

Table 6. Phenolic compounds quantified in almonds and almond skins (mg/100 g).

Compound		Almond	skin	References
		Flavonoids		
Flavan-3-ol	(+)-Catechin	0.1–36.6	0.69–18.4	[52,109–111,116,118,120–125]
	(-)catechin gallate	0.68–1.04		[52,126,127]
	Dihydrokaempferol	0.04–9.8	4.99–6.02	[111,118,126]
	Dihydroquercetin	0.51–1.60	n.d.–1.61	[110,111,118,128]
	(-)-Epicatechin	0.03–26.6	0.13–11.0	[52,58,109–111,118,120–123,128,129]
	Epicatechin gallate	1.34–2.60		[52,126,127]
	Gallocatechin gallate	n.d.–0.104		[58]
	(-)epigallocatechin gallate	1.04–1.60		[52]
	(-)epigallocatechin	8.07–8.87		[52]
	(-)gallocatechin	1.17–3.26		[52]
Flavanone	Eriodictyol	n.d.–0.46	n.d.–0.78	[58,110,111,118,120,126,128]
	Eridictyol-7-O-glucoside	n.d.–0.49	0.04–3.38	[58,110,111,118,120]
	Naringenin	0.01–9.74	0.03–20.6	[58,110,111,116,118,120,126,127,129]
	Naringenin-7-O-glucoside	0–5.88	0.04–14.3	[58,110,111,118,120,126,127,129]
Flavonol	Isorhamnetin	0.005–3.20	0.40–4.55	[58,110,111,118,120,124–130]
	Isorhamnetin-3-O-galactoside	0.30–0.92		[111,118]
	Isorhamnetin-3-O-glucoside	n.d.–14.9	0.20–16.9	[58,110,120,126,127,129]
	Isorhamnetin-3-O-rutinoside	n.d.–74.1	0.53–75.7	[58,110,111,118,120,126–129]
	Kaempferol	n.d.–0.49	0.01–1.25	[110,111,116,118,120,123–126,128–130]

Table 6. *Cont.*

Compound		Almond	skin	References
	Kaempferol-3-O-galactoside	n.d.–2.17	0.72–1.15	[111,118,126]
	Kaempferol-3-O-glucoside	n.d.–3.77	n.d.-39.0	[58,110,111,118,120,126,128,129]
	Kaempferol-3-O-rutinoside	n.d.–23.3	0.10–23.9	[58,109–111,118,120,126–129]
	Quercetin	n.d.–3.58	0.03–0.70	[110,111,118,120,123–126,128–130]
	Quercetin-3-O-galactoside	0.24–1.37	n.d.-1.34	[109–111,118,120,126]
	Quercetin-3-O-glucoside	0.04–0.16	n.d.-0.90	[109–111,118,120,129]
	Quercetin-3-O-rutinoside	n.d.–1.66	n.d.-41.2	[58,111,118,120,126,127,129]
	Phenolic acids/aldehydes			
Hydroxybenzoic acid	p-Hydroxybenzoic acid	n.d.–1.23	0.03–1.90	[58,110,111,116,118,120,126,129,131]
	Gallic acid	0.05–1.61	n.d.–1.61	[58,123,131]
	Protocatechuic acid	n.d.–6.19	0.04–4.46	[58,110,111,116,118,120,131]
	Vanillic acid	n.d.–0.30	0.01–5.81	[58,110,111,116,118,120,124,125,130,131]
	Ellagic acid	n.d.–0.135		[116]
Hydroxybenzoic aldehyde	Protocatechuic aldehyde	2.52–5.77	0.25–2.17	[110,111,118]
Hydroxycinnamic acid	Chlorogenic acid	n.d.-2.29	0.17–9.57	[58,110,111,116,118,120,123]
	Caffeic acid	0.11–3.21	n.d.–3.21	[116,123,125]
	o-Coumaric acid	0.22–0.69		[120,123–125,130]
	p-Coumaric acid	0.01–0.59	n.d.–0.37	[58,110,111,116,118]
	Ferulic acid	0.02–2.15		[116,125]
	Proanthocyanidins			
	A-type trimers		0.16–0.53	[111,118]
	Procyanidin B1	1.69–7.28		[111,118,121],
	Procyanidin B2	0.03–8.30	0.23–3.39	[110,111,118,122]
	Procyanidin B3	0.19–0.45		[111,118,121,124,127]
	Procyanidin B3+B1		0.30–2.96	[110,111,118,121,122,124,125]
	Procyanidin B5	n.d.-0.43	0.23–1.51	[110,111,118,121,122,124]
	Procyanidin B7	0.28–1.43	0.37–2.47	[110,111,118,121,122,124]
	Procyanidin C1		0.11–2.55	[110,121,122,124]
	Tannins			
	PAC dimers	4.00–18.7		[110,122,124]
	PAC trimers	2.70–14.0		[122,124]
	PAC 4–6 mers	7.00–51.4		[122,124]
	PAC 7–10 mers	9.60–52.0		[122,124]
	PAC polymers	43.9–121		[122,124]

6.7.1. Tannins

Proanthocyanidins are mixtures of oligomers and polymers of flavan-3-ol linked through carbon bonds, mainly C4 → C8. Tannins are divided into two groups, hydrolysable tannins and condensed tannins or proanthocyanidins (PAC). Hydrolysable tannins are derived from gallic acid and include gallotannins and ellagitannins. PACs are mixtures of oligomers and polymers of flavan-3-ol. Depending on the interflavan carbon–carbon bond, PACs could be A-type or B-type; depending on the degree of polymerization (DP), they are known as oligomeric (≤10) and polymeric proanthocyanidins (>10). In almond, most PACs are polymeric [130]. In addition, flavan-3-ol composition should be considered to determine PACs. The intrinsic complexity and diversity of almond proanthocyanidins, as well as a lack of available standards, pose analytical challenges [107].

Information about tannins is very limited and only a few studies quantify these compounds [81,107], even though PACs are the most abundant polyphenols in almond kernel, followed by flavonoids and phenolic acids. No references have been found about the presence of soluble tannins. Almond proanthocyanidins consist mainly of epicatechin and catechin, with lesser amounts of epiafzelechin. In the opinion of Bolling et al. [81,107], the cis–transconfiguration, A-/B-type ratios and flavan-3-ol types of almond PACs have not been adequately characterized.

PACs consisting exclusively of epicatechin are procyanidins (PCs). PACs containing epiafzelechin as subunits are named propelargonidins (PPs). When subunits are epigallocatechin, they are named prodelphinidins (PDs) [124].

Gu et al. [124] presented the concentrations of monomers, dimers and trimers separately because these low-molecular weight PACs oligomers (DP ≤ 3) could be absorbed intact in the gastrointestinal tract; meanwhile, PACs with DP > 3 appear not to be absorbed directly from the gastrointestinal lumen.

Table 6 summarizes the variability in tannins found in the reviewed papers published.

6.7.2. Flavonoids

At least 25 different flavonoids have been identified in almonds. Anthocyanidins, flavan-3-ols, flavonols, flavanones and a biflavone have been identified in almond, almond skins, or almond blanch water [107].

Most flavonoids of almond kernel appear exclusively in the skin, while non-flavonoids phenols appear in the seed. Flavonoids of almond skin act as phytoalexins, protecting dry seeds against bacteria, fungi and other environmental stressors [132]. The flavonoids group can be divided in seven categories that include flavonols, flavones, isoflavones, flavanols, flavanones, anthocyanins and dihydrochalcones.

Flavonols were the most abundant flavonoid class in almond and include isorhamnetin-3-O-glucoside, isorhamnetin-3-O-rutinoside, kaempferol-3-O-glucoside, quercetin-3-O-glucoside, quercetin-3-O-rutinoside, galactosides and rutinosides [107]. They have been identified, but rarely quantified (Table 6).

Flavonoid composition in plants is influenced by genetic factors and environmental conditions, excluding exposure to fungi and bacteria, parasites, climate and UV light [81,112,118].

6.7.3. Phenolic Acids

The presence of phenolic acids is associated with astringency, discoloration and inhibition of enzymatic activity and antioxidant properties. Phenolic acids identified in almond kernel have been caffeic, p-coumaric, ferulic, sinapic, syringic, vinyl, gallic, protocatechuic and p-hydroxybenzoic, which are fundamentally derivated from benzoic acid or cinnamic acid.

Senter and Horvat [131] found that protocatechuic acid is the dominant phenolic acid in the edible part of almond kernel, followed by p-hydroxybenzoic acid and vinyl acid. Wijeratne et al. [108] observed that the total quantity of free phenolic acids was 163 mg/100 g in the skin, while total quantities of esterified phenolic acids obtained from skin, shell and whole seed were 2796, 1671 and 400 mg/100 g, respectively. Monagas et al. [111] compared the polyphenol concentration in the almond skin of Spanish and American almonds (Table 6).

Bolling [107] reported the almond skin content of protocatechuic acids. The hydroxycinnamic acids, chlorogenic acid, ferulic acid, caffeic acid, sinapic acid and two diferulates have been identified in almond skin, although only chlorogenic acid, caffeic acid and p-coumaric acid have been quantitated.

6.7.4. Stilbene

Among stilbenes, resveratrol dominates, acting as a phytoalexin. Recently, Xie and Bolling [25] characterized stilbenes in Californian almond cultivars, concretely resveratrol-3-β-glucoside, in concentrations of 7.19 to 8.52 mg/100 g almonds. Similar to other polyphenols, stilbenes were concentrated in skins.

7. Volatile Compounds

Seventeen aroma compounds were detected in raw almonds [133], including six aldehydes, two ketones, two nitrogen-containing compounds, one sulfur-containing compound, two acids, one furanone and three unknown compounds. Six of these compounds were quantitated in raw almonds, where vanillin with 830 ng/g was the most abundant and acetic acid (137 ng/g) and nonanal (72 ng/g) were found in high abundance.

Ojeda-Amador et al. [35] analyzed volatile compounds which are related to sensory notes, such as fruit/banana (hexanol), oily/green–sweet (hexanal), fruity (pentanol) and bitter almonds (benzaldehyde). The most important family found in all the varieties studied was that of aldehydes (1.35–7.52 mg/kg). Benzaldehyde was the main aldehyde (52–74% of total), followed by hexanal (0–10%).

Alcohols were the second major family, accounting for 14% to 30% of the total volatiles. Hexanol was the main contributor and was most abundant in "Marcona" (1.89 mg/kg). Acids (mainly acetic acid), hydrocarbons, ketones and terpenes showed close concentrations to each other, indicating about 0.30 mg/kg for each family.

8. Conclusions

Almond kernel contains a considerable amount of good-quality proteins, mainly globulins, essential minerals and fiber with a low content in sugars, in addition to many phytochemicals with potential health benefits. The presence of large variability in nutritive compounds has been reported, although most pre- and postharvest factors may have a significant effect on their content. However deeper studies about drying, blanching, storage or roasting processes and genetic, agricultural and environmental conditions are necessary to clarify their influence on the quality and quantity of almond phytochemicals.

As regards the bibliography consulted, practically no work has been found focused on the study of the phenotypic correlations that occur between the different components of the almond.

The complexity in the phytochemical composition makes the use of standard methods for extracting and quantifying almond phytochemicals difficult. Non-conventional extraction techniques are gaining major interest, especially methods based on microwave, supercritical fluids and ultrasound, combined with well-known and safe solvents such as ethanol, water and ethanol–water mixtures. Other methods such as sonication and hydrolysis are barely cited in scientific papers.

In another direction, more studies are needed to understand the impact of almond processing on protein and AA digestibility. Furthermore, increasing efforts to establish a new method for assessing protein quality, based on the Digestible Indispensable Amino Acid Score (DIAAS) system, are necessary.

The valorization of non-lipid compounds from almond has been scarcely treated in the scientific literature. Most papers focus on compounds identification and quantification and rarely on industrial extraction methods, as opposed to oil extraction. However, the nutritional composition of the non-lipid fraction of almond kernel makes the by-products obtained in the oil extraction process interesting candidates for food applications, to be used as a source of protein, fiber and minerals.

Author Contributions: Conceptualization, J.M.R., M.Á.-O. and J.E.P.; methodology, J.M.R., A.P.-G. and, A.R.; formal analysis, J.M.R., A.P.-G., A.R.; data curation, J.M.R. and M.Á.-O.; writing—original draft preparation, J.M.R.; writing—review and editing, M.Á.-O., A.R. and J.E.P.; supervision, J.E.P. All authors have read and agreed to the published version of the manuscript.

Funding: This research received no external funding.

Conflicts of Interest: The authors declare no conflict of interest.

References

1. FAOSTAT. Available online: http://www.fao.org/faostat/en/#data/QC (accessed on 5 November 2020).
2. Felipe, A. *El Almendro. I. El Material Vegetal*; Mira Editores: Zaragoza, Spain, 2000.
3. Salazar, D.; Melgarejo, P. *El Cultivo del Almendro*; Mundi-Prensa: Madrid, Spain, 2002.
4. Barreca, D.; Nabavi, S.M.; Sureda, A.; Rasekhian, M.; Raciti, R.; Silva, A.S.; Annunziata, G.; Arnone, A.; Tenore, G.C.; Süntar, İ.; et al. Almonds (*Prunus dulcis* Mill. D. A. webb): A source of nutrients and health-promoting compounds. *Nutrients* **2020**, *12*, 672. [CrossRef] [PubMed]
5. Creedon, A.C.; Hung, E.S.; Berry, S.E.; Whelan, K. Nuts and their effect on gut microbiota, gut function and symptoms in adults: A systematic review and meta-analysis of randomised controlled trials. *Nutrients* **2020**, *12*, 2347. [CrossRef] [PubMed]
6. Siddiqui, I.; Husain, Q.; Azam, A. Exploring the antioxidant effects of peptides from almond proteins using PAni-Ag-GONC conjugated trypsin by improving enzyme stability & applications. *Int. J. Biol. Macromol.* **2020**, *158*, 150–158. [CrossRef] [PubMed]
7. Prats, M.S. Caracterización Quimiométrica de Diez Variedades de Almendra Cultivadas en Diferentes Localidades. Ph.D. Thesis, University of Alicante, Alicante, Spain, 2000.
8. Yada, S.; Lapsley, P.; Huang, G. A review of composition studies of cultivated almonds: Macronutrients and micronutrients. *J. Food Compos. Anal.* **2011**, *24*, 469–480. [CrossRef]
9. Yada, S.; Huang, G.; Lapsley, K. Natural variability in the nutrient composition of California-grown almonds. *J. Food Compos. Anal.* **2013**, *30*, 80–85. [CrossRef]
10. Kodad, O.; Estopañán, G.; Juan, T.; Alonso, J.M.; Espiau, M.T.; Rafel Socias i Company. Oil content, fatty acid composition and tocopherol concetration in the spanish almond genebank collection. *Sci. Horicult.* **2014**, *177*, 99–107. [CrossRef]
11. Rabadán, A.; Álvarez-Ortí, M.; Gómez, R.; de Miguel, C.; Pardo, J.E. Influence of genotype and crop year in the chemometrics of almond and pistachio oils. *J. Sci. Food Agric.* **2017**, *98*, 2402–2410. [CrossRef]
12. Rabadán, A.; Pardo, J.E.; Gómez, R.; Álvarez-Ortí, M. Influence of temperature in the extraction of nut oils by means of screw pressing. *LWT* **2018**, *93*, 354–361. [CrossRef]
13. Roncero, J.M.; Álvarez-Ortí, M.; Pardo-Giménez, A.; Gómez, R.; Rabadán, A.; Pardo, J.E. Almond virgin oil: Parameters of regulated physicochemical quality and stability. *Riv. Ital. Sostanze Grasse* **2016**, *93*, 237–243.
14. Font i Forcada, C.; Kodad, O.; Juan, T.; Estopañan, G.; Rafel Socias i Company. Genetic variability and pollen effect on the transmission of the chemical components of the almond kernel. *Span. J. Agric. Res.* **2011**, *9*, 781–789. [CrossRef]
15. Saura, F.; Cañellas, J.; Soler, L. *La Almendra. Composición, Variedades, Desarrollo y Maduración*; Instituto Nacional de Investigaciones Agrarias (INIA): Madrid, Spain, 1988.
16. Casares, R.L.H. Chemical data on the raises of Málaga. *An. Bromatol.* **1952**, *4*, 411–419.
17. Hall, A.M. The nutritive value of fresh and roasted, California-grown Nonpareil almonds. *J. Agric. Food Chem.* **1958**, *6*, 377–382. [CrossRef]
18. Esteban, R.M. *Estudio Comparativo de la Calidad Nutritiva de Variedades de Almendra del S.E. y N.E. español*; Instituto Nacional de Investigación y Tecnología Agraria y Alimentaria (INIA): Madrid, Spain, 1985.
19. Salvo, F.A. Composizionee dell'olio di mandorle. Nota III: Variazione di alcuni parametri chimici e chimico-ficici durante la conservazione. *Riv. Ital. Sostanze Grasse* **1986**, *63*, 37–40.
20. García, C.G. Major fatty acid composition of 19 almond cultivars of different origins. A chemometric approach. *J. Agric. Food Chem.* **1996**, *44*, 1751–1755. [CrossRef]
21. Martín, M.L. Comparative study on the triglyceride composition of almond kernel oil. A new basis for cultivar chemometric characterization. *J. Agric. Food Chem.* **1999**, *47*, 3688–3692. [CrossRef]

22. Kodad, O. Criterios de Selección y de Evaluación de Nuevas Obtenciones Autocompatibles en un Programa de Mejora Genética del Almendro (*Prunus amygdalus* Batsch). Ph.D. Thesis, University of Lleida, Lleida, Spain, 2006.
23. Barile, D.; Coïsson, J.D.; Arlorio, M.; Rinaldi, M. Identification of production area of Ossolano Italian cheese with chemometric complex approach. *Food Control.* **2006**, *17*, 197–206. [CrossRef]
24. Costa, J.; Mafra, I.; Carrapatoso, I.; Oliveira, M.B. Almond allergens: Molecular characterization, detection, and clinical relevance. *J. Agric. Food Chem.* **2012**, *60*, 1337–1349. [CrossRef]
25. Xie, L.; Bolling, B.W. Characterisation of stilbenes in California almonds (*Prunus dulcis*) by UHPLC–MS. *Food Chem.* **2014**, *148*, 300–306. [CrossRef]
26. Vanamala, J. Food systems approach to cancer prevention. *Crit. Rev. Food Sci. Nutr.* **2017**, *57*, 2573–2588. [CrossRef]
27. Bolling, B. Polyphenol content and antioxidant activity of California almonds depend on cultivar and harvest year. *Food Chem.* **2010**, *122*, 819–825. [CrossRef]
28. Zhu, Y.W. Lipophilic antioxidant content of almonds (*Prunus dulcis*): A regional and varietal study. *J. Food Compos. Anal.* **2015**, *39*, 120–127. [CrossRef]
29. Csakvari, A.C.; Lupitu, A.; Bungău, S.; Gîtea, M.A.; Gîtea, D.; Ţiţ, D.M.; Copolovici, D. Fatty acids profile and antioxidant activity of almond oils obtained from six Romanian varieties. *Farmacia* **2019**, *67*, 882–887. [CrossRef]
30. Maestri, D.; Martínez, M.; Bodoira, R.; Rossi, Y.; Oviedo, A.; Pierantozzi, P. Variability in almond oil chemical traits from traditional cultivars and native genetic resources from Argentina. *Food Chem.* **2015**, *170*, 55–61. [CrossRef]
31. Sena-Moreno, E.; Pardo, J.E.; Pardo-Giménez, A.; Gómez, R.; Alvarez-Ortí, M. Differences in Oils from Nuts Extracted by Means of Two Pressure Systems. *Int. J. Food Prop.* **2016**, *19*, 2750–2760. [CrossRef]
32. Roncero, J.M.; Alvarez-Ortí, M.; Pardo-Giménez, A.; Gómez, R.; Rabadán, A.; Pardo, J.E. Virgin almond oil: Extraction methods and composition. *Grasas Aceites* **2016**, *67*, e143. [CrossRef]
33. Martínez, M.L.; Bordón, M.G.; Bodoira, R.M.; Penci, M.C.; Ribotta, P.D.; Maestri, D.M. Walnut and almond oil screw-press extraction at industrial scale: Effects of process parameters on oil yield and quality. *Grasas Aceites* **2017**, *68*, e216. [CrossRef]
34. Pardo-Giménez, A.; Carrasco, J.; Roncero, J.M.; Álvarez-Ortí, M.; Zied, D.C.; Pardo, J.E. Recycling of the biomass waste defatted almond meal as a novel nutritional supplementation for cultivated edible mushrooms. *Acta Sci. Agron.* **2018**, *40*, e39341. [CrossRef]
35. Ojeda-Amador, R.; Fregapane, G.; Salvador, M. Chemical Characterization of Virgin Almond and Hazelnut Oils and Their By-Products. *Eur. J. Lipid Sci. Technol.* **2019**, *121*, 1–10. [CrossRef]
36. Barreira, J.C.M.; Nunes, M.A.; da Silva, B.V.; Pimentel, F.B.; Costa, A.S.G.; Alvarez-Ortí, M.; Pardo, J.E.; Oliveira, M.B.P.P. Almond cold-pressed oil by-product as ingredient for cookies with potential health benefits: Chemical and sensory evaluation. *Food Sci. Hum. Wellness* **2019**, *8*, 292–298. [CrossRef]
37. Rabadán, A.; Álvarez-Ortí, M.; Martínez, E.; Pardo-Giménez, A.; Zied, D.C.; Pardo, J.E. Effect of replacing traditional ingredients for oils and flours from nuts and seeds on the characteristics and consumer preferences of lamb meat burgers. *LWT* **2021**, *136*, 110307. [CrossRef]
38. Sathe, S.K. Solubilization, electrophoretic characterization and in vitro digestibility of almond (*Prunus amygdalus*) proteins. *J. Food Biochem.* **1993**, *16*, 249–264. [CrossRef]
39. Kodad, O.; Mamouni, A.; Lahlo, M.; Rafel Socias i Company. Contenido en aceite y proteína y de los caracteres físicos del fruto y de la pepita del almendro en las condiciones climáticas mediterráneas. *ITEA Inf. Tec. Econ. Agrar.* **2011**, *107*, 300–314.
40. Ozcan, M.; Ünver, A.; Erkan, E.; Arslan, D. Characteristics of some almond kernel and oils. *Sci. Hortic.* **2011**, *127*, 330–333. [CrossRef]
41. Drogoudi, P.D.; Pantelidis, G.; Bacchetta, L.; Giorgio, D.; Duval, H.; Metzidakis, I. Protein and mineral nutrient contents in kernels from 72 sweet almond cultivars and accessions grown in France, Greece and Italy. *Int. J. Food Sci. Nutr.* **2012**, *64*, 202–209. [CrossRef] [PubMed]
42. Kodad, O.; Estopañán, G.; Juan, T.; Socias i Compani, R. Protein content and oil composition of almond from moroccan seedlings: Genetic diversity, oil quality and geographical origin. *J. Am. Oil Chem. Soc.* **2013**, *90*, 243–252. [CrossRef]

43. USDA Food Database U.S. Department of Agriculture. Available online: https://fdc.nal.usda.gov/download-datasets.html (accessed on 21 February 2019).
44. Osborne, T.B.; Campbell, G.F. Conglutin and vitellin. *J. Am. Chem. Soc.* **1986**, *18*, 609–623. [CrossRef]
45. Ahrens, S.; Venkatachalam, M.; Mistry, A.M.; Lapsley, K.; Sathe, S.K. Almond (*Prunus dulcis* L.) protein quality. *Plant Foods Hum. Nutr.* **2005**, *60*, 123–128. [CrossRef]
46. Barreira, J.; Casal, S.; Ferreira, I.; Peres, A.M.; Pereira, J. Supervised chemical pattern recognition in almond (*Prunus dulcis*) portuguese PDO cultivars: PCA- and LDA-Based triennial study. *J. Agric. Food Chem.* **2012**, *60*, 9697–9704. [CrossRef]
47. Askin, M.; Baltab, M.; Tekintasc, F.; Kazankayab, A.; Balta, F. Fatty acid composition affected by kernel weight in almond (*Prunus dulcis* Mill.) genetic resources. *J. Food Compos. Anal.* **2007**, *20*, 7–12. [CrossRef]
48. Simsek, M.; Gulsoy, E.; Yavic, A.; Arikan, B.; Yildirim, Y.; Olmez, N.; Erdogmus, B.; Boguc, F. Fatty acid, mineral and proximate compositions of various genotypes and commercial cultivars of sweet almond from the same ecological conditions. *Appl. Ecol. Environ. Res.* **2018**, *16*, 2957–2971. [CrossRef]
49. Agumbiade, S.O. Evaluation of some nutritional characteristics of inidian almond (*Prunus amygdalus*) nut. *Pak. J. Nutr.* **2006**, *5*, 316–318. [CrossRef]
50. Chung, K.H.; Shin, K.O.; Hwang, H.J.; Choi, K.S. Chemical composition of nuts and seeds sold in Korea. *Nutr. Res. Pract.* **2013**, *7*, 82–88. [CrossRef] [PubMed]
51. Moodley, R.; Kindness, A.; Jonnalagadda, S. Elemental composition and chemical characteristics of five edible nuts (almond, Brazil, pecan, macadamia and walnut) consumed in Southern Africa. *J. Environ. Sci. Health Part B* **2007**, *58*, 585–591. [CrossRef] [PubMed]
52. Čolić, S.D.; Bakić, I.V.; Zagorac, D.C.; Natić, M.M.; Smailagić, A.T.; Pergal, M.V.; Pešić, M.B.; Milinčić, D.D.; Rabrenović, B.B.; Akšić, M.M. Chemical Fingerprint and Kernel Quality Assessment in Different Grafting Combinations of Almond Under Stress Condition. *Sci. Hortic.* **2020**, *275*, 109705. [CrossRef]
53. Akpambang, V.; Amoo, I.; Izuagie, A. Comparative compositional analysis on two varieties of melon (*Colocynthis citrullus* and *Cucumeropsis edulis*) and a variety of almond (*Prunus amygdalus*). *Res. J. Agric. Biol. Sci.* **2008**, *4*, 639–642. [CrossRef]
54. Aslantas, R.; Guleryuz, M.; Turan, M. Some chemical contents of selected almond (*Prunus amygdalus* Batsch) types. *Cah. Options Méditerr.* **2001**, *56*, 347–350.
55. Kazantzis, I.; Nanos, G.; Stavroulakis, G. Effect of harvest time and storage conditions on almond kernel oil and sugar composition. *J. Sci. Food Agric.* **2003**, *83*, 354–359. [CrossRef]
56. Barreira, J.; Pereira, J.; Oliveira, M.; Ferreira, I. Sugars profiles of different chestnut (*Castanea sativa* Mill.) and almond (*Prunus dulcis*) cultivars by HPLC-RI. *Plant Foods Hum. Nutr.* **2010**, *65*, 38–43. [CrossRef]
57. Sánchez-Bel, P.; Egea, I.; Martınez-Madrid, M.; Flores, B.; Romojaro, F. Influence of irrigation and organic/inorganic fertilization on chemical quality of almond (*Prunus amygdalus* cv. Guara). *J. Agric. Food Chem.* **2008**, *56*, 10056–10062. [CrossRef]
58. Oliveira, I.; Meyer, A.; Afonso, S.; Aires, A.; Goufo, P. Phenolic and fatty acid profiles, α-tocopherol and sucrose contents, and antioxidant capacities of understudied Portuguese almond cultivars. *J. Food Biochem.* **2019**, *43*, e12887. [CrossRef]
59. Ruggeri, S.; Cappelloni, M.; Gambelli, L.; Carnovale, E. Chemical composition and nutritive value of nuts grown in Italy. *Ital. J. Food Sci.* **1998**, *10*, 243–252.
60. Mandalari, G.; Nueno-Palop, C.; Bisignano, G.; Wickham, M.S.; Narbad, A. Potential prebiotic properties of almond (*Amygdalus communis* L.) seeds. *Appl. Environ. Microbiol.* **2008**, *74*, 4264–4270. [CrossRef] [PubMed]
61. Font i Forcada, C.; Fernández i Martí, A.; Rafel Socias i Company. Mapping quantitative trait loci for kernel composition in almond. *BMC Genet.* **2012**, *13*, 47. [CrossRef] [PubMed]
62. Venkatachalam, M.; Sathe, S. Chemical composition of selected edible nut seeds. *J. Agric. Food Chem.* **2006**, *54*, 4705–4714. [CrossRef]
63. Barbera, G.; Di Marco, L.; La Mantia, T.; Schirra, M. Effect of rootstock on productive and qualitative response of two almond varieties. *Acta Hortic.* **1994**, *373*, 129–134. [CrossRef]
64. Sathe, S.K. Biochemical characterization of amandin, the major storage protein in almond (*Prunus dulcis* L.). *J. Agric. Food Chem.* **2002**, *50*, 4333–4341. [CrossRef]
65. Grané-Teruel, N.; Luna, M.; Berenguer, V. Statistical comparative study of free amino acid HPLC data from a selected almond set. *Food Chem.* **1999**, *65*, 23–28. [CrossRef]

66. Amrein, T.; Lukac, H.; Andres, L.; Perren, R.; Escher, F.; Amado, R. Acrylamide in roasted almonds and hazelnuts. *J. Agric. Food Chem.* **2005**, *53*, 7819–7825. [CrossRef]
67. Serón, L.A.; Garrigós, E.S.; Berenguer, V.; Grané-Teruel, N. Characterisation of 19 almond cultivars on the basis of their free amino acids composition. *Food Chem.* **1998**, *61*, 455–459. [CrossRef]
68. House, J.D.; Hill, K.; Neufeld, J.; Franczyk, A.; Nosworthy, M.G. Determination of the protein quality of almonds (*Prunus dulcis* L.) as assessed by in vitro and in vivo methodologies. *Food Sci. Nutr.* **2019**, *7*, 2932–2938. [CrossRef]
69. Nanos, G.; Kazantzis, I.; Kefalas, P.; Petrakis, C.; Stavroulakis, G. Irrigation and harvest time affect almond kernel quality and composition. *Sci. Hortic.* **2002**, *96*, 249–256. [CrossRef]
70. Rabadán, A.; Álvarez-Ortí, M.; Gómez, R.; Pardo-Giménez, A.; Pardo, J.E. Suitability of Spanish almond cultivars for the industrial production of almond oil and defatted flour. *Sci. Hortic.* **2017**, *225*, 539–546. [CrossRef]
71. Arrázola, G. Análisis de Glucósidos Cianogénicos en Variedades de Almendros: Implicaciones en la Mejora Genética. Ph.D. Thesis, University of Alicante, Alicante, Spain, 2002.
72. Lipan, L.; Martín-Palomo, M.J.; Sánchez-Rodríguez, L.; Cano-Lamadrid, M.; Sendra, E.; Hernańdez, F.; Burlo, F.; Vázquez-Araújo, L.; Andreu, L.; Carbonell-Barrachina, A.A. Almond fruit quality can be improved by means of deficit irrigation strategies. *Agric. Water Manag.* **2019**, *217*, 236–242. [CrossRef]
73. FOSS. *Fiber Determination Using the Fibertec I&M Systems*; Application Note AN304; FOSS Tecator AB: Höganäs, Sweden, 2001.
74. Schirra, M.; Mulas, M.; Nieddu, G.; Virdis, F. Mineral content in Texas almonds during fruit growth and ripening. *Acta Hortic.* **1994**, *373*, 207–214. [CrossRef]
75. Martínez, M.L.; Penci, M.C.; Marin, M.A.; Ribotta, P.D.; Maestri, D.M. Screw press extraction of almond (*Prunus dulcis* Miller): Oil recovery and oxidative stability. *J. Food Eng.* **2013**, *119*, 40–45. [CrossRef]
76. Sudhakar, P.; Priyanka, K.; Peter, A.E.; Sandeep, B.V.; Rajeswari, M.; Rao, B.; Sujatha, P. A study on the proximate composition and nutritive value of local tree almonds, *Prunus amygdalus*. *Ann. Plant Sci.* **2018**, *7*, 2363–2372. [CrossRef]
77. Jing, Z.; Cheng, J.; Guo, C.; Wang, X. Seed traits, nutrient elements and assessment of genetic diversity for almond (*Amygdalus* spp.) endangered to China as revealed using SRAP markers. *Biochem. Syst. Ecol.* **2013**, *49*, 51–57. [CrossRef]
78. Ozcan, M.M. Determination of the mineral compositions of some selected oil-bearing seeds and kernels using Inductively Coupled Plasma Atomic Emission Spectrometry (ICP-AES). *Grasas Aceites* **2006**, *57*, 211–218. [CrossRef]
79. Cabrera, C.; Lloris, F.; Gimenez, R.; Olalla, M.; Lopez, M.C. Mineral content in legumes and nuts: Contribution to the Spanish dietary intake. *Sci. Total Environ.* **2003**, *308*, 1–14. [CrossRef]
80. Alasalvar, C.; Pelvan, E. Fat-soluble bioactives in nuts. *Eur. J. Lipid Sci. Technol.* **2011**, *113*, 943–949. [CrossRef]
81. Bolling, B.; Chen, C.Y.; McKay, D.; Blumberg, J.B. Tree nut phytochemicals: Composition, antioxidant capacity, bioactivity, impact factors. A systematic review of almonds, Brazils, cashews, hazelnuts, macadamias, pecans, pine nuts, pistachios and walnuts. *Nutr. Res. Rev.* **2011**, *24*, 244–275. [CrossRef] [PubMed]
82. Chen, C.Y.; Lapsley, K.; Blumberg, J. A nutrition and health perspective on almond. *J. Sci. Food Agric.* **2006**, *86*, 2245–2250. [CrossRef]
83. Stuetz, W.; Schlormann, W.; Glei, M. B-vitamins, carotenoids and α-/γ-tocopherol in raw and roasted nuts. *Food Chem.* **2017**, *221*, 222–227. [CrossRef] [PubMed]
84. Kornsteiner, M.W. Tocopherols and total phenolics in 10 different nut types. *Food Chem.* **2006**, *98*, 381–387. [CrossRef]
85. Madawala, S.K.; Dutta, P. Lipid components and oxidative status of selected specialty oils. *Grasas Aceites* **2012**, *63*, 143–151. [CrossRef]
86. Kodad, O.; Estopañán, G.; Juan, T.; Socias i Company, R. Tocopherol concentration in almond oil from Moroccan seedlings: Geographical origin and post-harvest implications. *J. Food Compos. Anal.* **2014**, *33*, 161–165. [CrossRef]
87. Ozcan, M.; Al Juhaimi, F.; Ghafoor, K.; Babiker, E. Characterization of physico-chemical and bioactive properties of oils of some important almond cultivars by cold press and soxhlet extraction. *J. Food Sci. Technol.* **2020**, *57*, 955–961. [CrossRef]

88. Rizzolo, A.; Baldo, C.; Polesello, A. Application of high-performance liquid chromatography to the analysis of niacin and biotin in Italian almond cultivars. *J. Chromatogr. A* **1991**, *553*, 187–192. [CrossRef]
89. Rafel Socias i Company; Kodad, O.; Alonso, J.M.; Font-Forcada, C. Fruit quality in almond: Physical aspects for breeding strategies. *Acta Hortic.* **2009**, *814*, 475–480. [CrossRef]
90. Cherif, A.; Belkacemi, K.; Kallel, H.; Angers, P.; Arul, J.; Boukhchina, S. Phytosterols, unsaturated fatty acid composition and accumulation in the almond kernel during harvesting period: Importance for development regulation. *Comptes Rendus Biol.* **2009**, *332*, 1069–1077. [CrossRef]
91. Azadmard, S.; Emami, S.H.; Hesari, J.; Peighambardoust, S.H.; Nemati, M. Nuts Composition and their Health Benefits. *World Acad. Sci. Eng. Technol.* **2011**, *81*, 508–512. [CrossRef]
92. Givianrad, M.; Saber-Tehrani, M.; Jafari, J.A. Chemical composition of oils from wild almond (*Prunus scoparia*) and wild pistachio (*Pistacia atlantica*). *Grasas y Aceites* **2013**, *64*, 77–84. [CrossRef]
93. Matthäus, B.; Özcan, M.M.; Juhaimi, F.; Adiamo, O.Q.; Alsawmahi, O.N.; Ghafoor, K.; Babiker, E. Effect of the Harvest Time on Oil Yield, Fatty Acid, Tocopherol and Sterol Contents of Developing Almond and Walnut Kernels. *J. Oleo Sci.* **2018**, *67*, 39–45. [CrossRef] [PubMed]
94. Fernandes, G.D.; Gómez-Coca, R.B.; Pérez-Camino, M.C.; Moreda, W.; Barrera-Arellano, D. Chemical Characterization of Major and Minor Compounds of Nut Oils: Almond, Hazelnut, and Pecan Nut. *J. Chem.* **2017**, 2609549. [CrossRef]
95. Miraliakbari, H.; Shahidi, F. Lipid class compositions, tocopherols and sterols of tree nut oils extracted with different solvents. *J. Food Lipids* **2008**, *15*, 81–96. [CrossRef]
96. Phillips, K.M.; Ruggio, D.; Ashraf-Khorassanij, M. Phytosterol composition of nuts and seeds commonly consumed in the United States. *J. Agric. Food Chem.* **2005**, *53*, 9436–9445. [CrossRef] [PubMed]
97. Chen, C.; Blumberg, J. In vitro activity of almond skin polyphenols for scavenging free radicals and inducing quinone reductase. *J. Agric. Food Chem.* **2008**, *56*, 4427–4434. [CrossRef]
98. Maguire, L.O.; Galvin, K.; O'Connor, T.O. Fatty acid profile, tocopherol, squalene and phytosterol content of walnuts, almonds, peanuts, hazelnuts and the macadamia nut. *Int. J. Food Sci. Nutr.* **2004**, *55*, 171–178. [CrossRef]
99. Alasalvar, C.S. *Tree Nuts: Composition, Phytochemicals and Health Effects*; CRC Press, Taylor and Francis Group: Boca Raton, FL, USA, 2008. [CrossRef]
100. Malisiova, F.H.S. Liposomal formulations from phospholipids of Greek almond oil. Properties and biological activity. *Verl. Z. Nat.* **2004**, *59*, 330–334. [CrossRef]
101. Boukhchina, S.; Sebai, K.; Cherif, A.; Kallel, H.; Mayer, P. Identification of glycerophospholipids in rapeseed, olive, almond, and sunflower oils by LC-MS and LC-MS. *Can. J. Chem.* **2004**, *82*, 1210–1215. [CrossRef]
102. Fang, F.H.T. Determination of sphingolipids in nuts and seeds by a single quadrupole liquid chromatography-mass spectrometry method. *J. Food Lipids* **2005**, *12*, 327–343. [CrossRef]
103. Song, S.; Cheong, L.; Wang, H.; Man, Q.; Pang, S.; Li, Y. Characterization of phospholipid profiles in six kinds of nut using HILIC-ESI-IT-TOF-MS system. *Food Chem.* **2018**, *240*, 1171–1178. [CrossRef] [PubMed]
104. Ogah, O.; Watkins, C.S.; Ubi, B.E.; Oraguzie, N.C. Phenolic compounds in Rosaceae fruit and nut crops. *J. Agric. Food Chem.* **2014**, *62*, 9369–9386. [CrossRef] [PubMed]
105. Lara, M.V.; Bonghi, C.; Famiani, F.; Vizzotto, G.; Walker, R.P.; Drincovich, M.F. Stone Fruit as Biofactories of Phytochemicals with Potential Roles in Human Nutrition and Health. *Front. Plant Sci.* **2020**, *11*, 562252. [CrossRef] [PubMed]
106. Bodoira, R.; Maestri, D. Phenolic Compounds from Nuts: Extraction, Chemical Profiles, and Bioactivity. *J. Agric. Food Chem.* **2020**, *68*, 927–942. [CrossRef]
107. Bolling, B. Almond Polyphenols: Methods of Analysis, Contribution to Food Quality, and Health Promotion. *Compr. Rev. Food Sci. Food Saf.* **2017**, *16*, 346–368. [CrossRef]
108. Wijeratne, S.; Abou-Zaid, M.; Shahidi, F. Antioxidant polyphenols in almond and its coproducts. *J. Agric. Food Chem.* **2006**, *54*, 312–318. [CrossRef]
109. Milbury, P.E.; Chen, C.Y.; Dolnikowski, G.G.; Blumberg, J.B. Determination of flavonoids and phenolics and their distribution in almonds. *J. Agric. Food Chem.* **2006**, *54*, 5027–5033. [CrossRef]
110. Bartolomé, B.; Monagas, M.; Garrido, I.; Gómez-Cordovés, C.; Martín-Álvarez, P.J.; Lebrón-Aguilar, R.; Urpí-Sardà, M.; Llorach, R.; Andrés-Lacueva, C. Almond (*Prunus dulcis* (Mill.) D.A. Webb) polyphenols: From chemical characterization to targeted analysis of phenolic metabolites in humans. *Arch. Biochem. Biophys.* **2010**, *501*, 124–133. [CrossRef]

111. Monagas, M.; Garrido, I.; Lebron-Aguilar, R.; Bartolome, B.; Gomez-Cordoves, C. Almond (*Prunus dulcis* (Mill.) D.A. Webb) Skins as a Potential Source of Bioactive Polyphenols. *J. Agric. Food Chem.* **2007**, *55*, 8498–8507. [CrossRef]
112. Barreira, J.C.; Ferreira, I.C.; Oliveira, M.B.; Pereira, J.A. Antioxidant activity and bioactive compounds of ten Portuguese regional and commercial almond cultivars. *Food Chem. Toxicol.* **2008**, *46*, 2230–2235. [CrossRef] [PubMed]
113. Hanine, H.; Zinelabidine, L.; Hssaini, H.; Nablousi, A.; Ennahli, S.; Latrache, H. Pomological and biochemical characterization of almond cultivars in Morocco. *Turk. J. Agric. Nat. Sci.* **2014**, *1*, 743–753.
114. Lombardi-Boccia, G.; Lucarini, M.; Lanzi, S.; Aguzzi, A.; Cappelloni, M. Nutrients and antioxidant molecules in yellow plums (*Prunus domestica* L.) from conventional and organic productions: A comparative study. *J. Agric. Food Chem.* **2004**, *52*, 90–94. [CrossRef] [PubMed]
115. Wu, X.; Beecher, G.; Holden, J.; Haytowitz, D.; Gebhardt, S.; Prior, R. Lipophilic and hydrophilic antioxidant capacities of common foods in the United States. *J. Agric. Food Chem.* **2004**, *52*, 4026–4037. [CrossRef]
116. Čolić, S.D.; Fotiric Aksic, M.M.; Lazarevic, K.B.; Zec, G.N.; Gasic, U.M.; Dabic Zagorac, D.C.; Natic, M.M. Fatty acid and phenolic profiles of almond grown in Serbia. *Food Chem.* **2017**, *234*, 455–463. [CrossRef]
117. Garrido, I.; Monagas, M.; Gómez-Cordovés, C.; Bartolomé, B. Extracción de antioxidantes a partir de subproductos del procesado de la almendra. *Grasas Aceites* **2007**, *58*, 130–135. [CrossRef]
118. Garrido, I.; Monagas, M.; Gómez-Cordovés, C.; Bartolomé, B. Polyphenols and antioxidant properties of almond skins: Influence of industrial processing. *J. Food Sci.* **2008**, *73*, 106–115. [CrossRef]
119. Ma, X.; Zhou, X.-Y.; Qiang, Q.-Q.; Zhang, Z.-Q. Ultrasound assisted extraction and preliminary purification of proanthocyanidins and chlorogenic acid from almond (*Prunus dulcis*) skin. *J. Sep. Sci.* **2014**, *37*, 1834–1841. [CrossRef]
120. Lin, J.T.; Liu, S.C.; Hu, C.C.; Shyu, Y.S.; Hsu, C.Y.; Yang, D.J. Effects of roasting temperature and duration on fatty acid composition, phenolic composition, Maillard reaction degree and antioxidant attribute of almond (*Prunus dulcis*) kernel. *Food Chem.* **2016**, *190*, 520–528. [CrossRef]
121. Bittner, K.; Rzeppa, S.; Humpf, H.U. Distribution and quantification of flavan-3-ols and procyanidins with low degree of polymerization in nuts, cereals, and legumes. *J. Agric. Food Chem.* **2013**, *61*, 9148–9154. [CrossRef]
122. Xie, L.; Roto, A.V.; Bolling, B.W. Characterization of ellagitannins, gallotannins, and bound proanthocyanidins from California almond (*Prunus dulcis*) varieties. *J. Agric. Food Chem.* **2012**, *60*, 12151–12156. [CrossRef] [PubMed]
123. Yıldırım, A.N.; San, B.; Koyuncu, F.; Yıldırım, F. Variability of phenolics, α-tocopherol and amygdalin contents of selected almond (*Prunus amygdalus* Batsch.) genotypes. *J. Food Agric. Environ.* **2010**, *8*, 50–53. [CrossRef]
124. Gu, L.; Kelm, M.A.; Hammerstone, J.F.; Beecher, G.; Holden, J.; Haytowitz, D.; Gebhardt, S.P.; Ronald, L. Concentrations of Proanthocyanidins in Common Foods and Estimations of Normal Consumption. *J. Nutr.* **2004**, *134*, 613–617. [CrossRef] [PubMed]
125. Amarowicz, R.; Troszynska, A.; Shahidi, F. Antioxidant activity of almond seed extract and its fractions. *J. Food Lipids* **2005**, *12*, 344–358. [CrossRef]
126. Bolling, B.W.; Dolnikowski, G.; Blumberg, J.B.; Chen, C.Y. Quantification of almond skin polyphenols by liquid chromatography-mass spectrometry. *J. Food Sci.* **2009**, *74*, 326–332. [CrossRef] [PubMed]
127. Valdés, A.; Vidal, L.; Beltrán, A.; Canals, A.; Garrigós, M.C. Microwave-assisted extraction of phenolic compounds from almond skin byproducts (*Prunus amygdalus*): A multivariate analysis approach. *J. Agric. Food Chem.* **2015**, *63*, 5395–5402. [CrossRef]
128. Qureshi, M.N.; Numonov, S.; Abudurexiti, A.; Aisa, H.A. Phytochemical investigations and evaluation of antidiabetic potential of *Prunus dulcis* nuts. *LWT Food Sci. Technol.* **2016**, *66*, 311–317. [CrossRef]
129. Hughey, C.A.; Janusziewicz, R.; Minardi, C.S.; Phung, J.; Huffman, B.A.; Reyes, L.; Wilcox, B.E.; Prakash, A. Distribution of almond polyphenols in blanch water and skins as a function of blanching time and temperature. *Food Chem.* **2012**, *131*, 1165–1173. [CrossRef]
130. Gu, L.; Kelm, M.A.; Hammerstone, J.F.; Beecher, G.; Holden, J.; Haytowitz, D. Screening of foods containing proanthocyanidins and their structural characterization using LC-MS/MS and thiolytic degradation. *J. Agric. Food Chem.* **2003**, *51*, 7513–7521. [CrossRef]

131. Senter, S.D.; Horvat, R.J. Comparative GLC-MS analysis of phenolic acids of selected tree nuts. *J. Food Sci.* **1983**, *48*, 798–799. [CrossRef]
132. Koes, R.; Quattrocchio, F.; Mol, J. The flavonoid biosynthetic pathway in plants: Function and evolution. *BioEssays* **1994**, *16*, 123–132. [CrossRef]
133. Erten, E.; Cadwallader, K. Identification of predominant aroma components of raw, dry roasted and oil roasted almonds. *Food Chem.* **2017**, *217*, 244–253. [CrossRef] [PubMed]

MDPI
St. Alban-Anlage 66
4052 Basel
Switzerland
Tel. +41 61 683 77 34
Fax +41 61 302 89 18
www.mdpi.com

Foods Editorial Office
E-mail: foods@mdpi.com
www.mdpi.com/journal/foods

www.ingramcontent.com/pod-product-compliance
Lightning Source LLC
Chambersburg PA
CBHW062035220926
51850CB00010B/2562